Graph Paper

Deluxe

Gray Line Edition

Kimberlite
Kreations

Graph Paper Deluxe

Graph Paper Deluxe

Graph Paper Deluxe

Graph Paper Deluxe

Graph Paper Deluxe

Graph Paper Deluxe

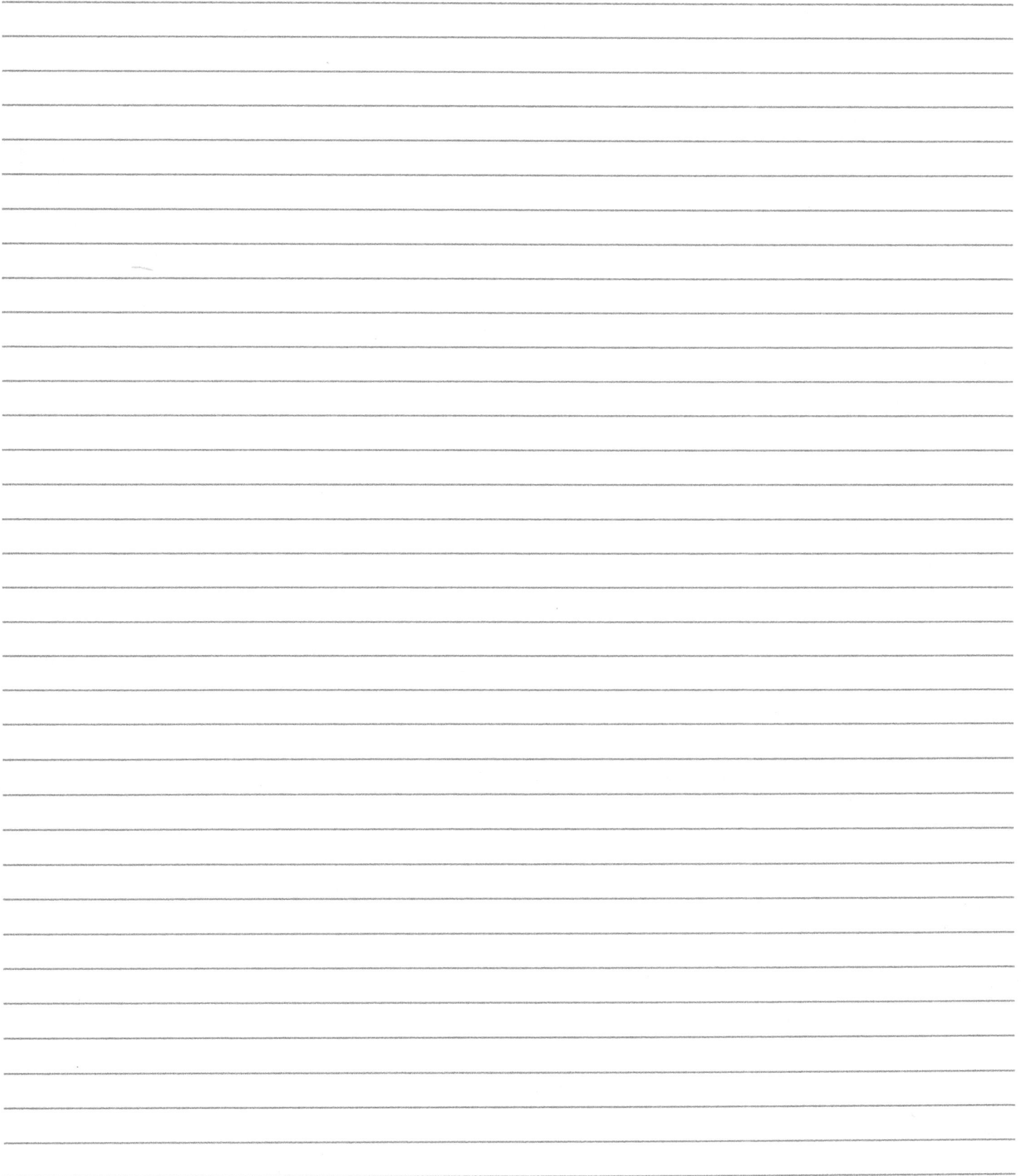

Gray Line Edition

Blessings and Eternal Love

Graph Paper Deluxe

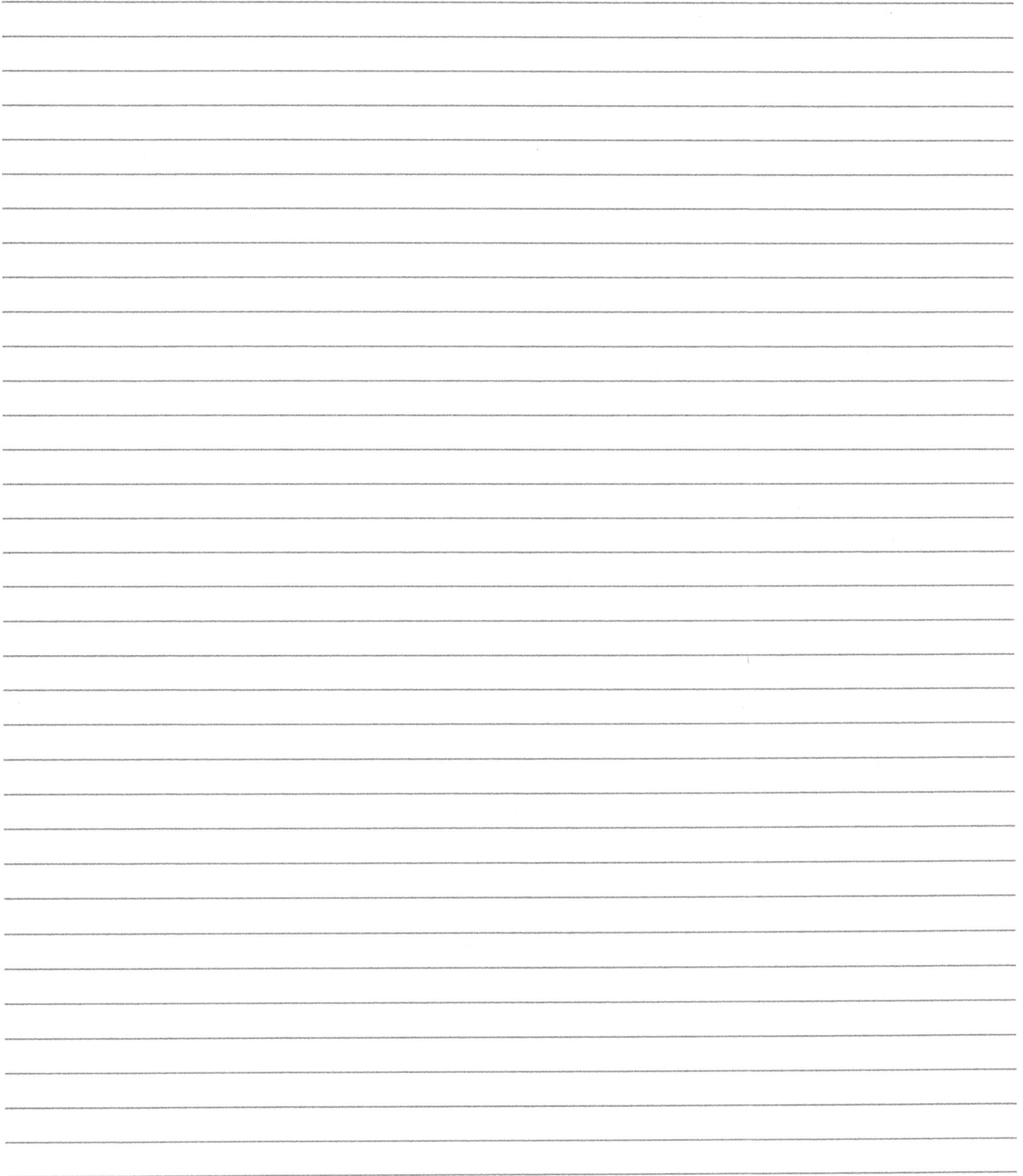

Gray Line Edition

Blessings and Eternal Love

Graph Paper Deluxe

Graph Paper Deluxe

Kimberlite Kreations

Gray Line Edition

Blessings and Eternal Love

Graph Paper Deluxe

Kimberlite Kreations

Gray Line Edition

Blessings and Eternal Love

Graph Paper Deluxe

Kimberlite Kreations

Graph Paper Deluxe

Kimberlite Kreations

Graph Paper Deluxe

Graph Paper Deluxe

Kimberlite Kreations

Graph Paper Deluxe

Kimberlite Kreations

Graph Paper Deluxe

Blessings and Eternal Love

Graph Paper Deluxe

Kimberlite Kreations

Gray Line Edition

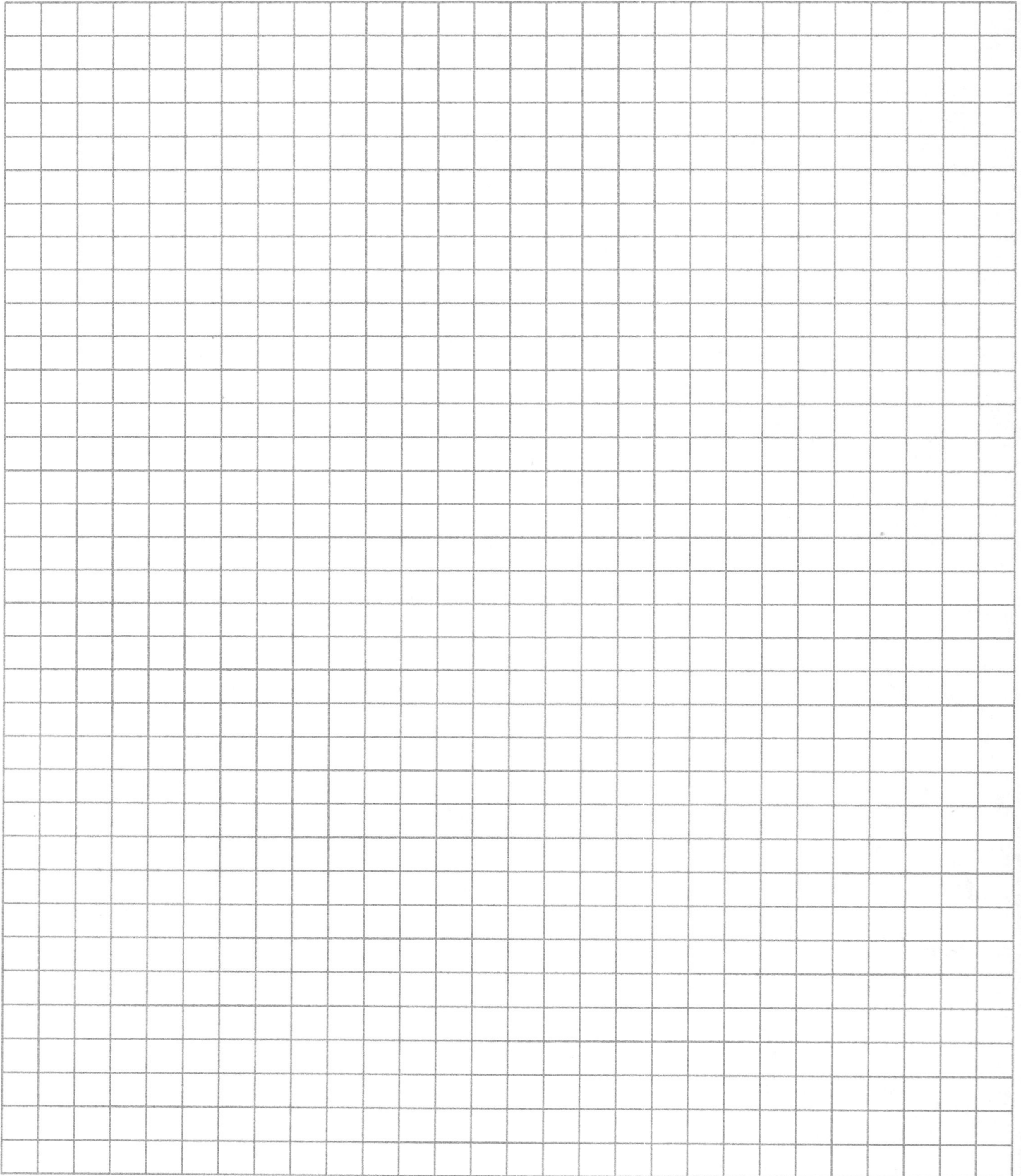

Blessings and Eternal Love

Graph Paper Deluxe

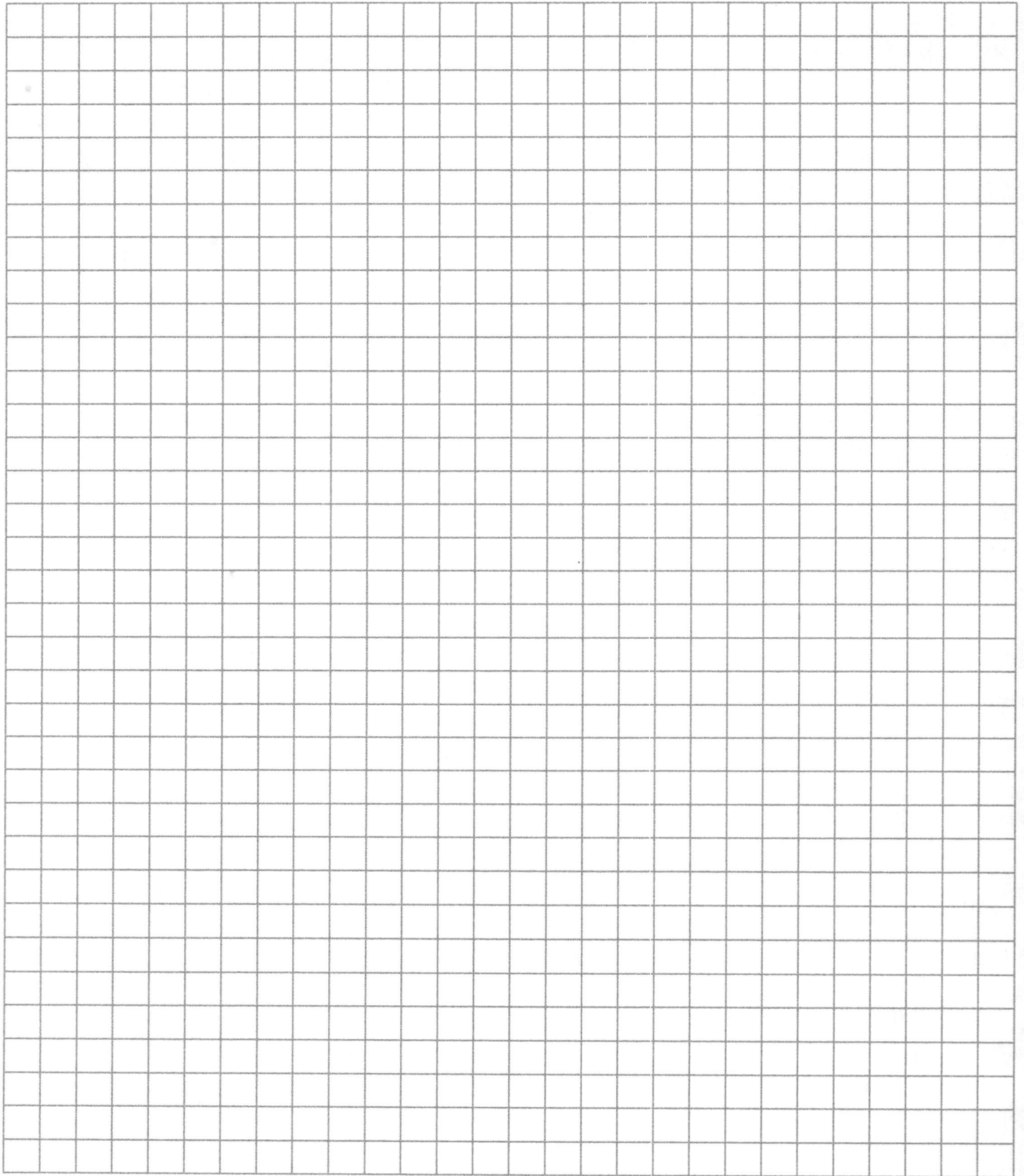

Graph Paper Deluxe

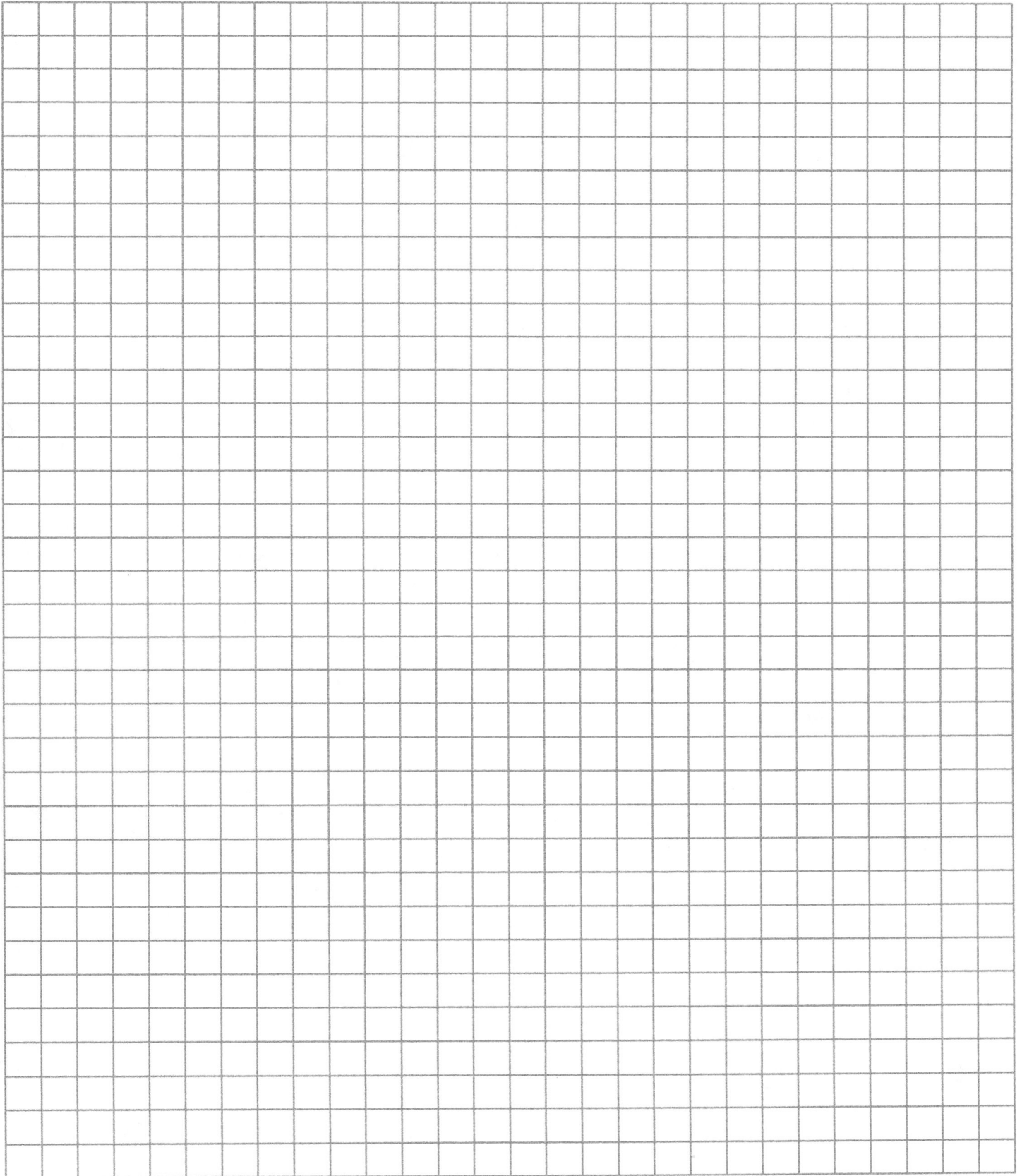

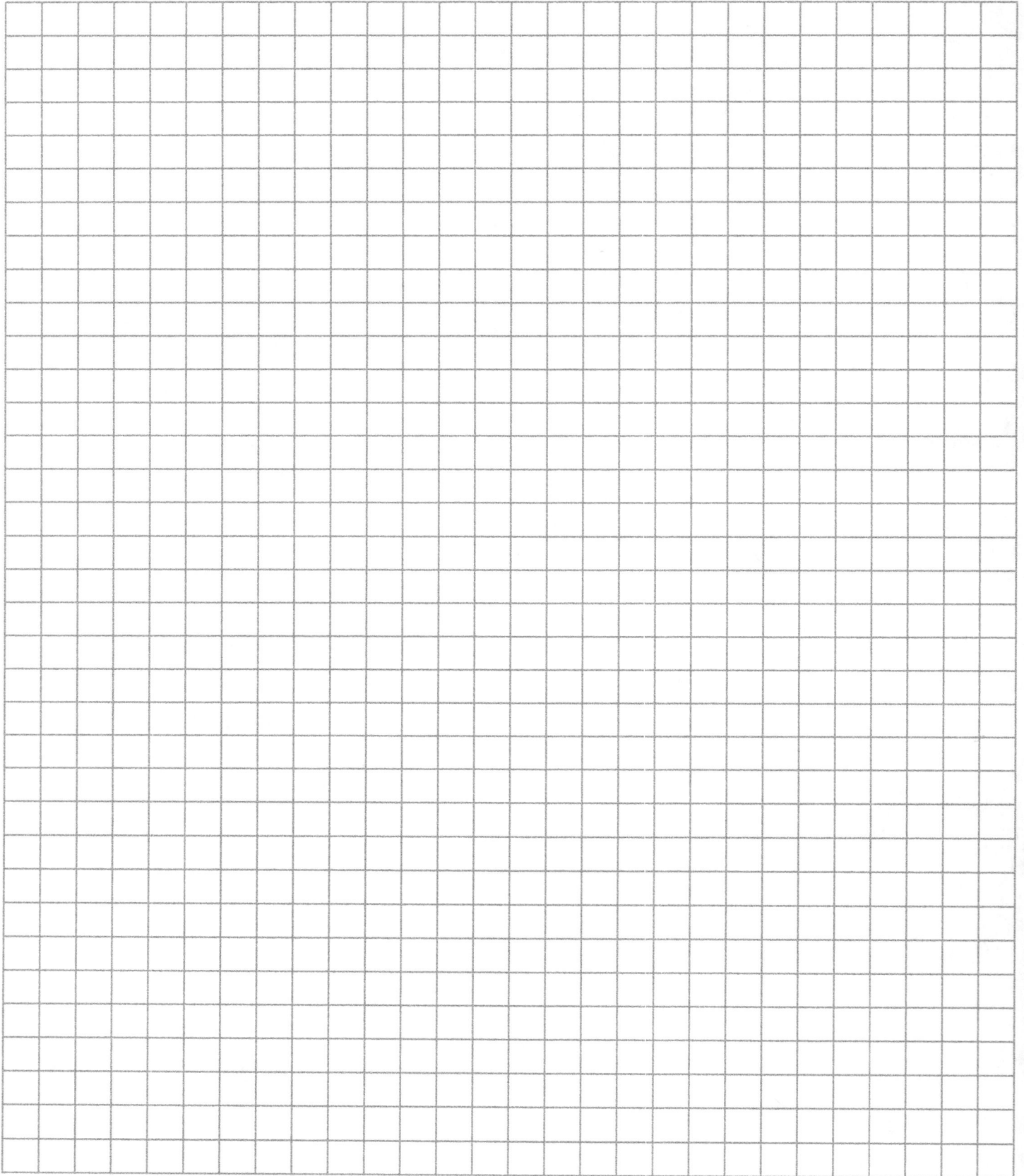

Graph Paper Deluxe

Gray Line Edition

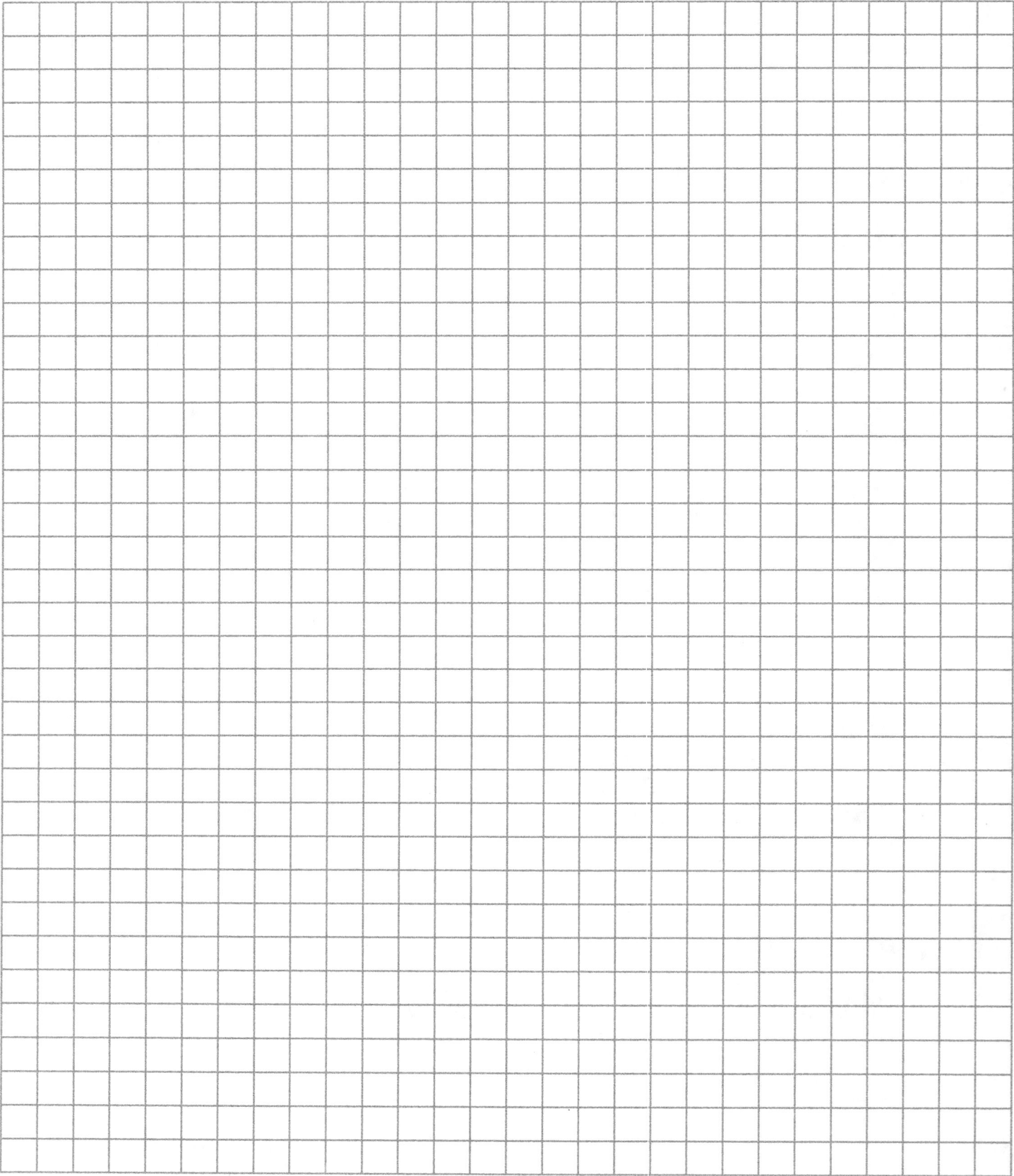

Blessings and Eternal Love

Graph Paper Deluxe

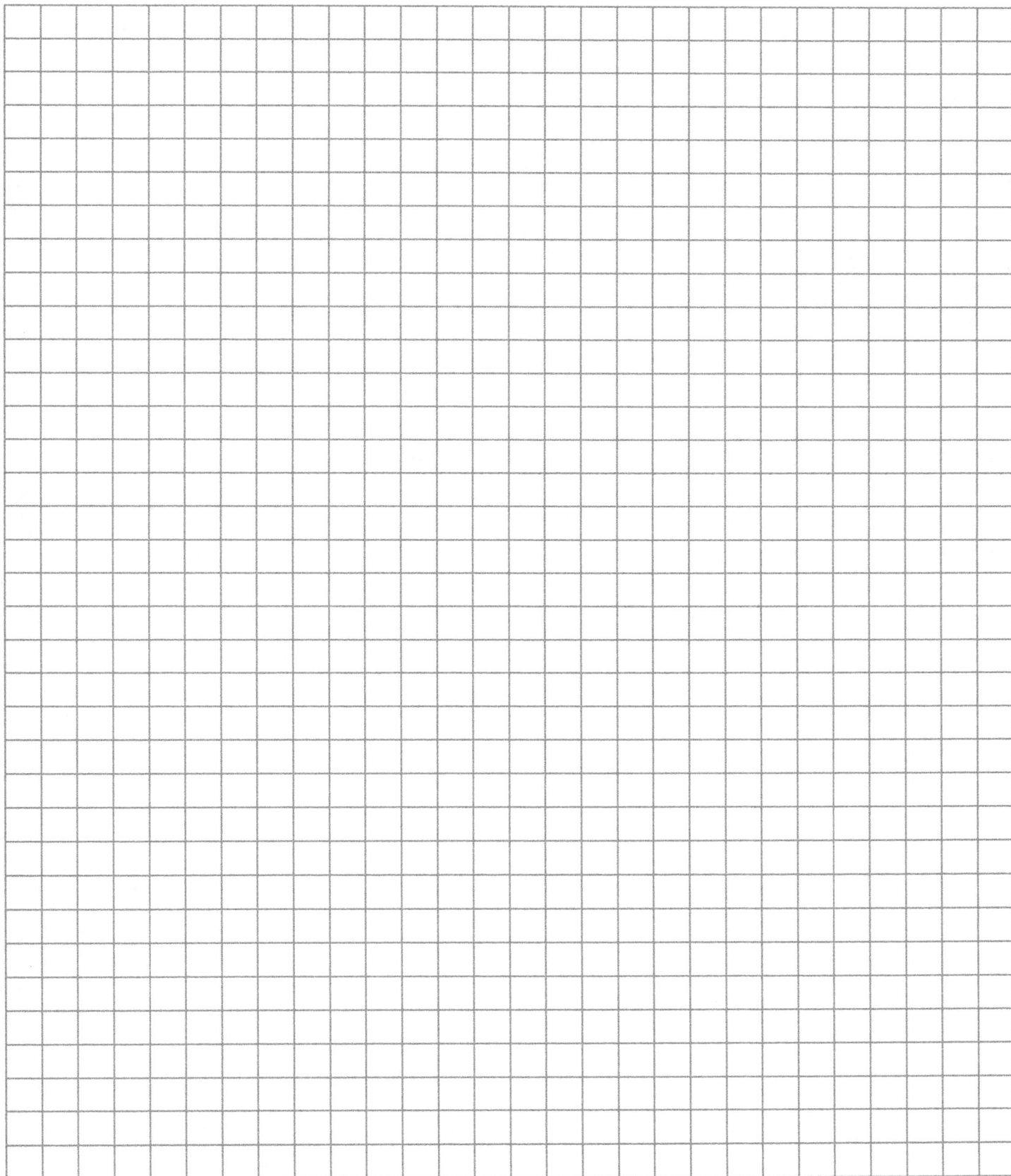

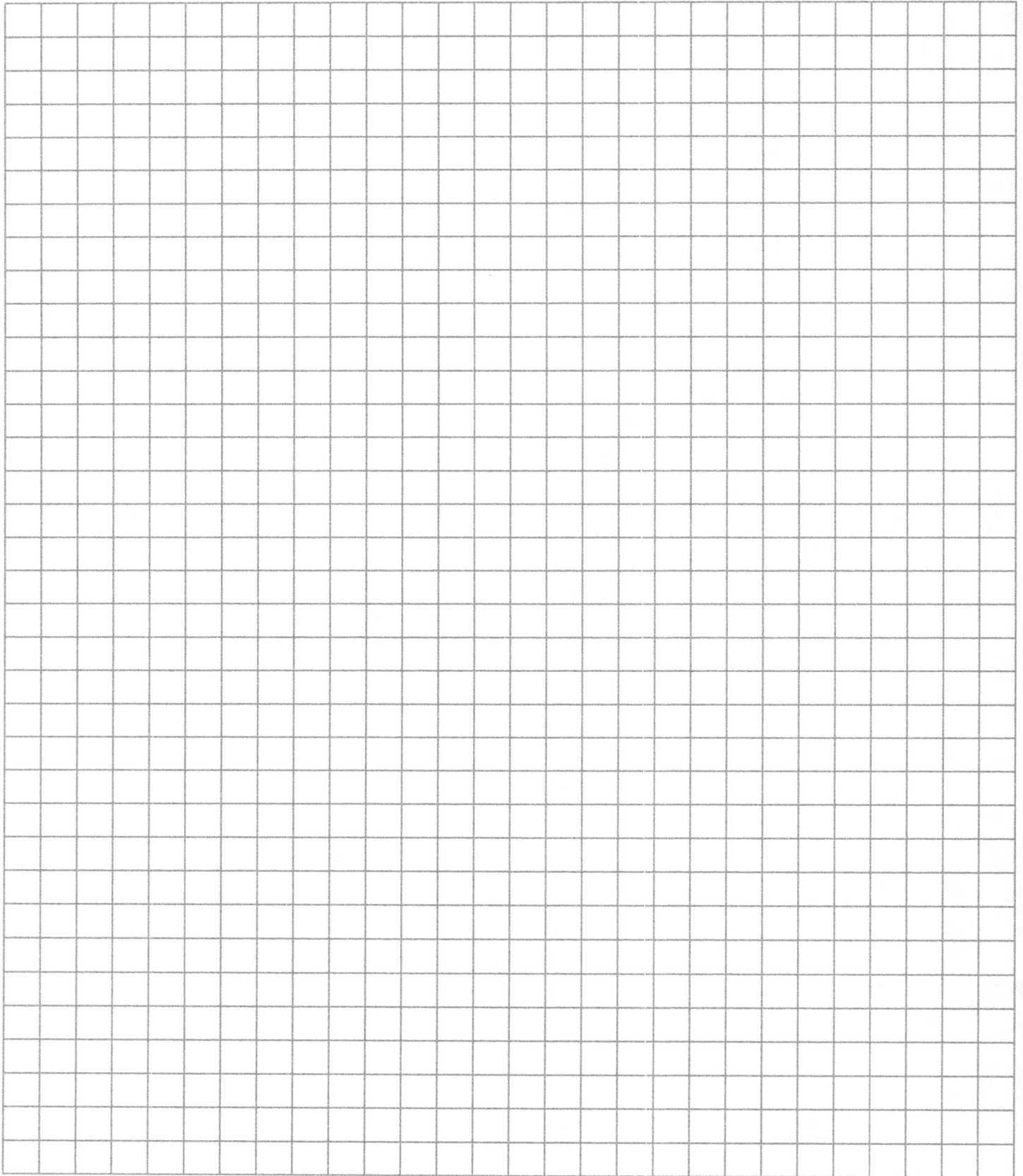

Graph Paper Deluxe

Gray Line Edition

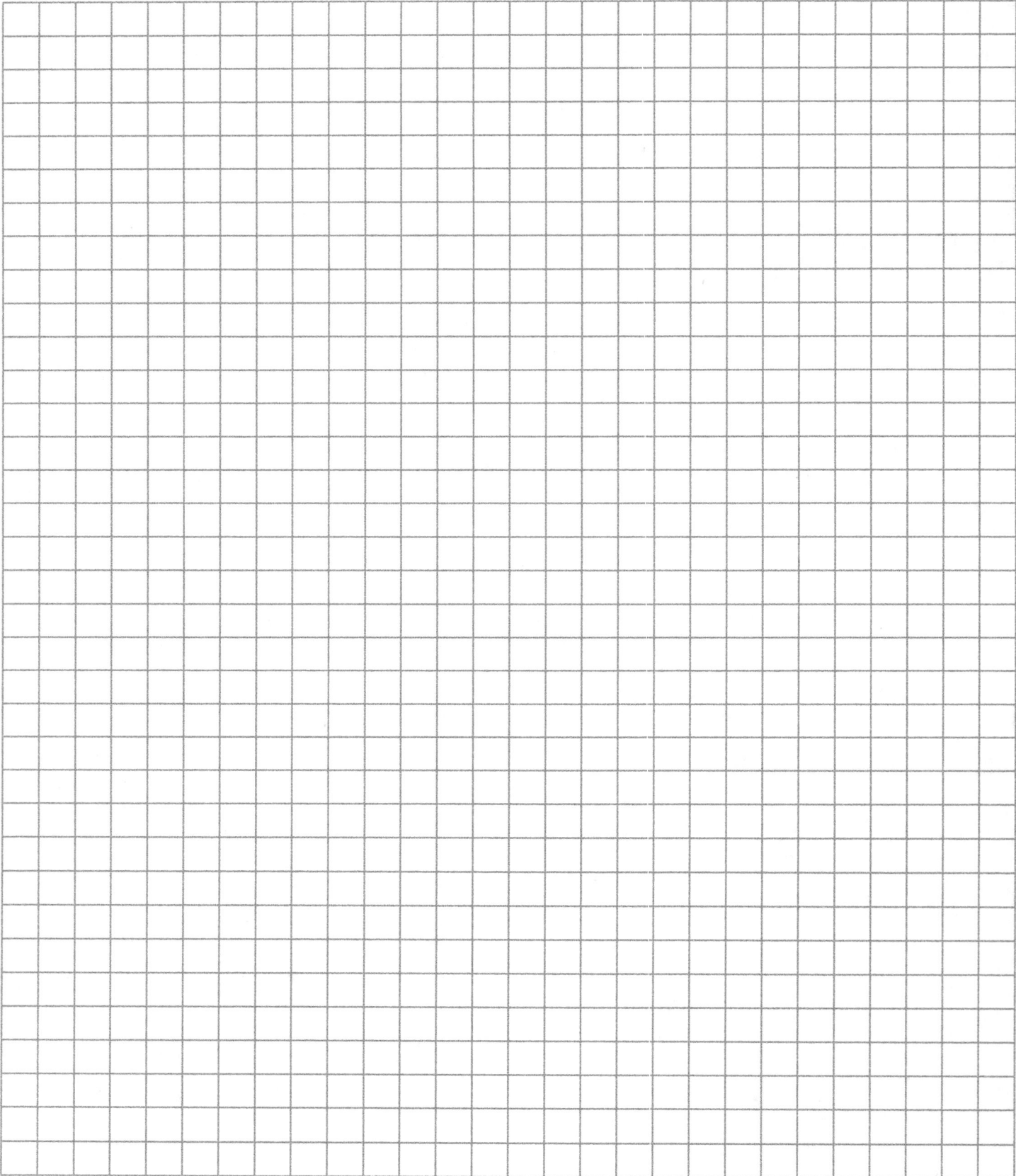

Blessings and Eternal Love

Graph Paper Deluxe

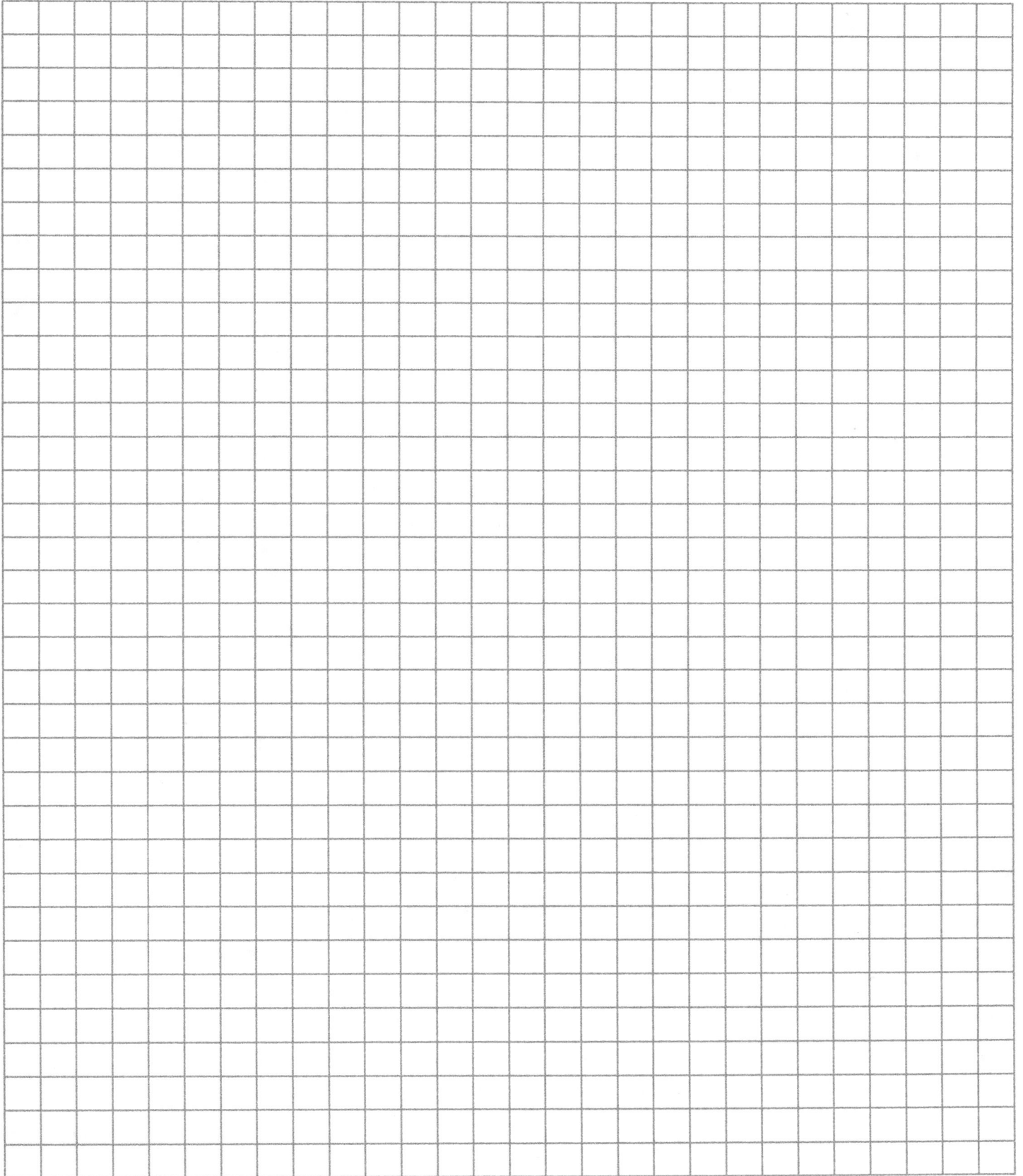

Graph Paper Deluxe

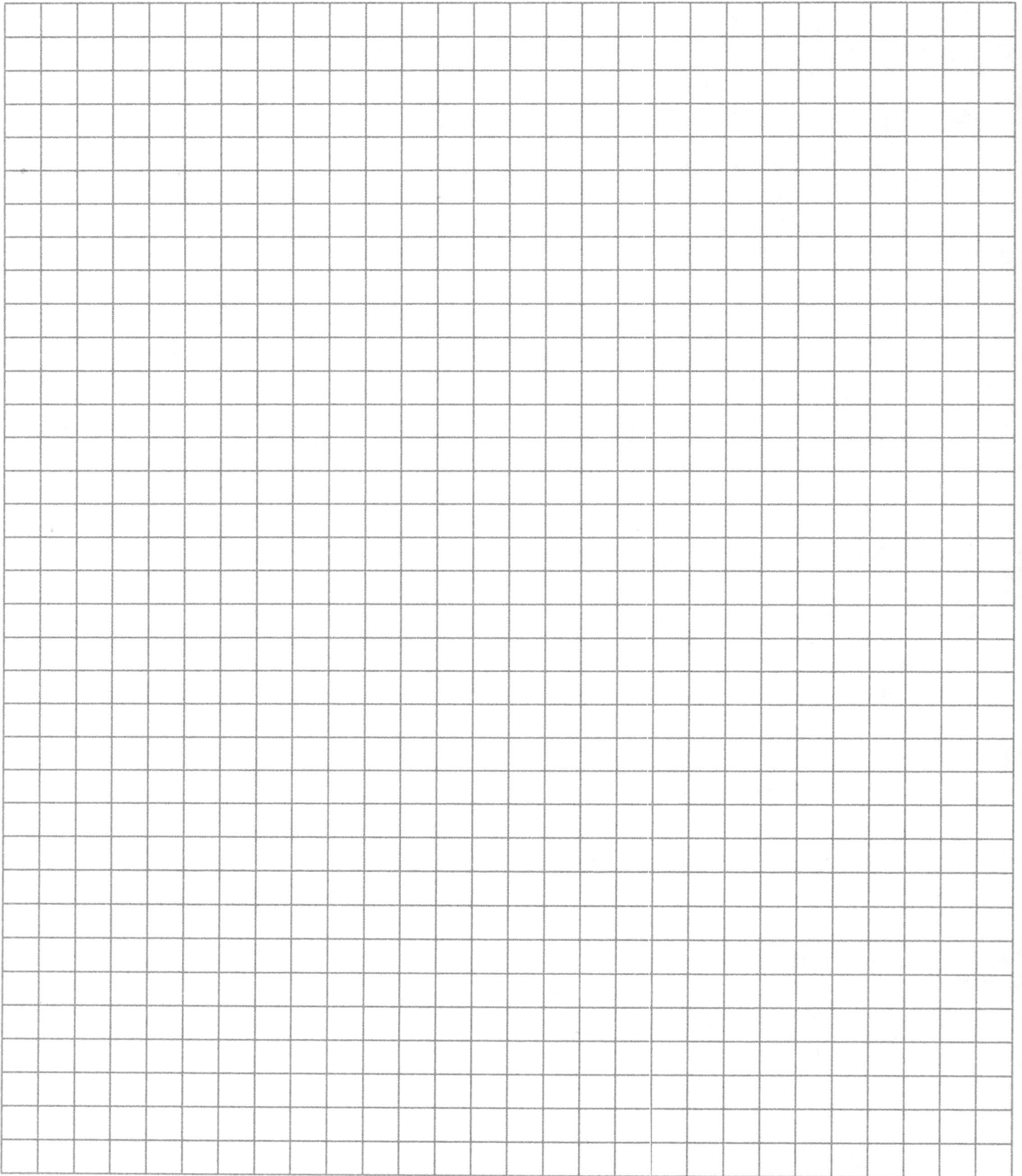

Graph Paper Deluxe

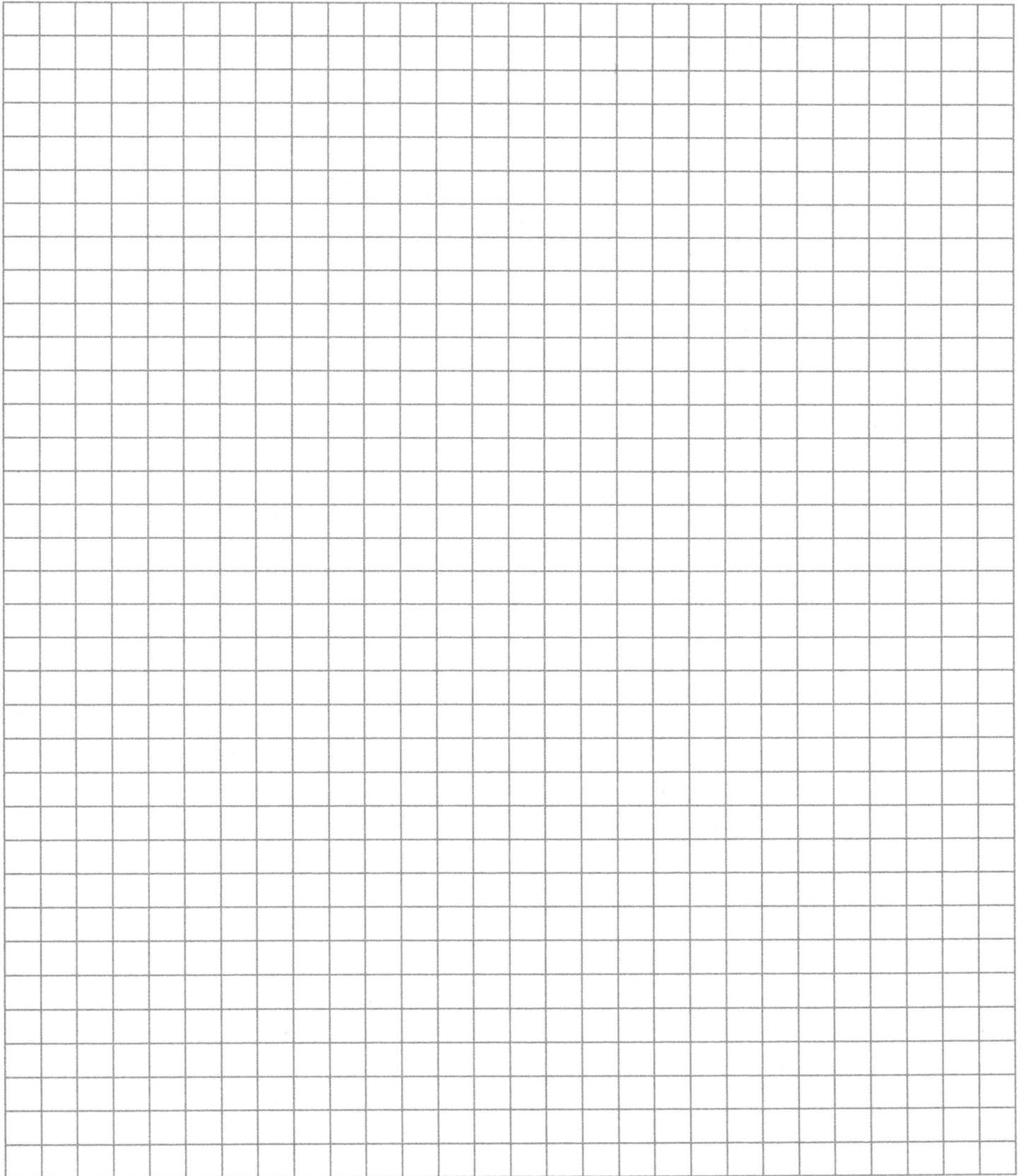

Gray Line Edition

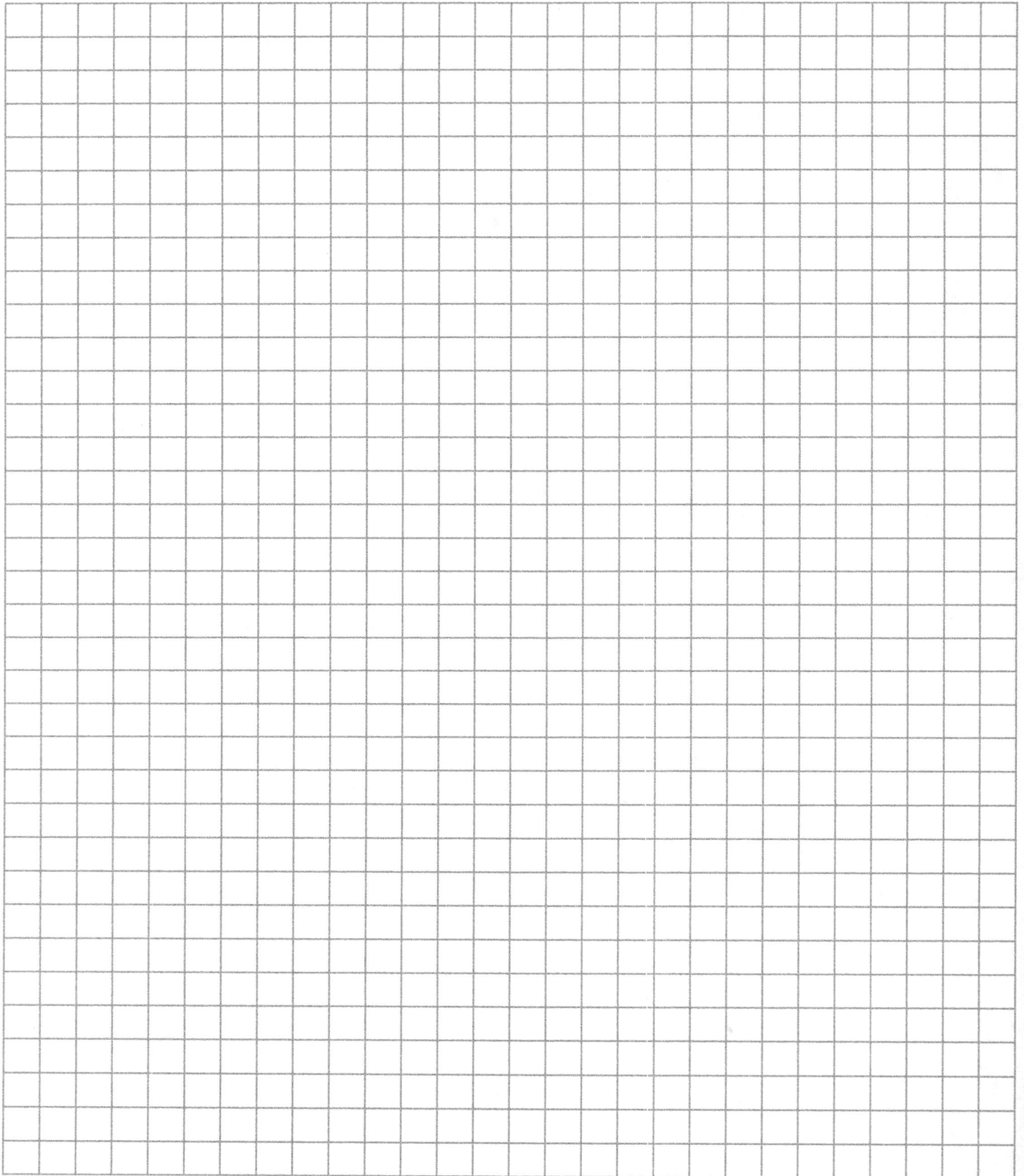

Blessings and Eternal Love

Graph Paper Deluxe

Gray Line Edition

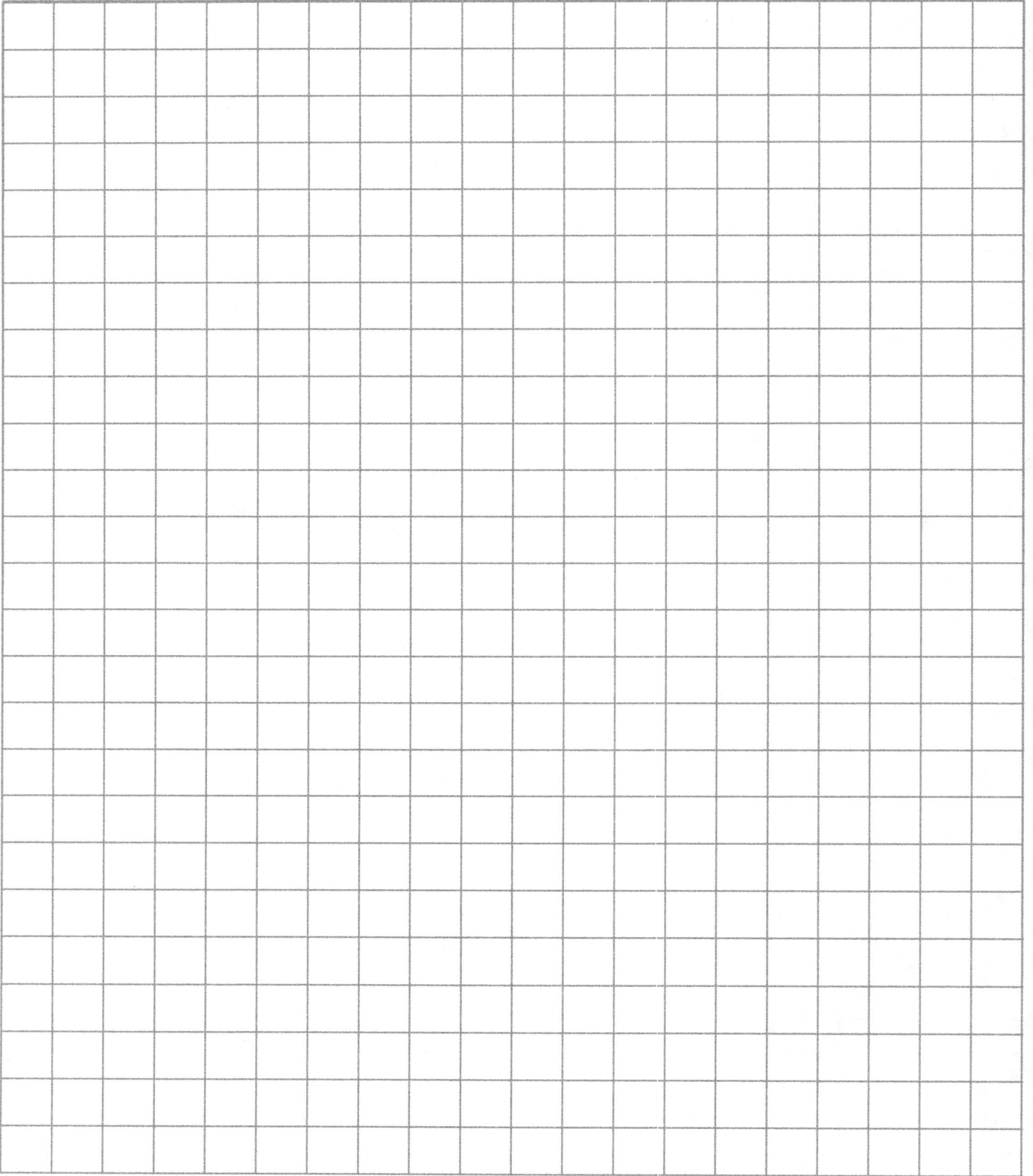

Blessings and Eternal Love

Graph Paper Deluxe

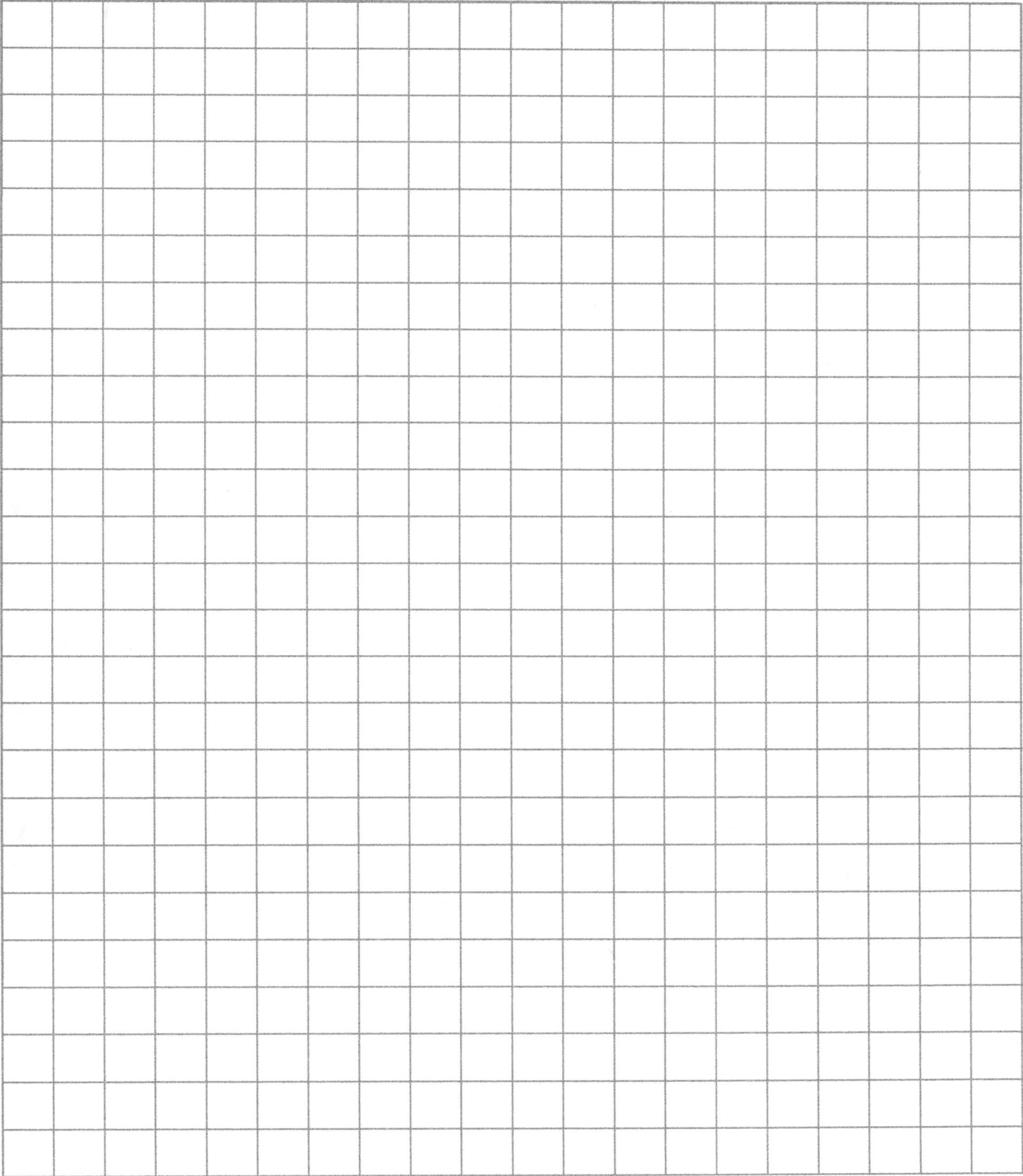

Gray Line Edition

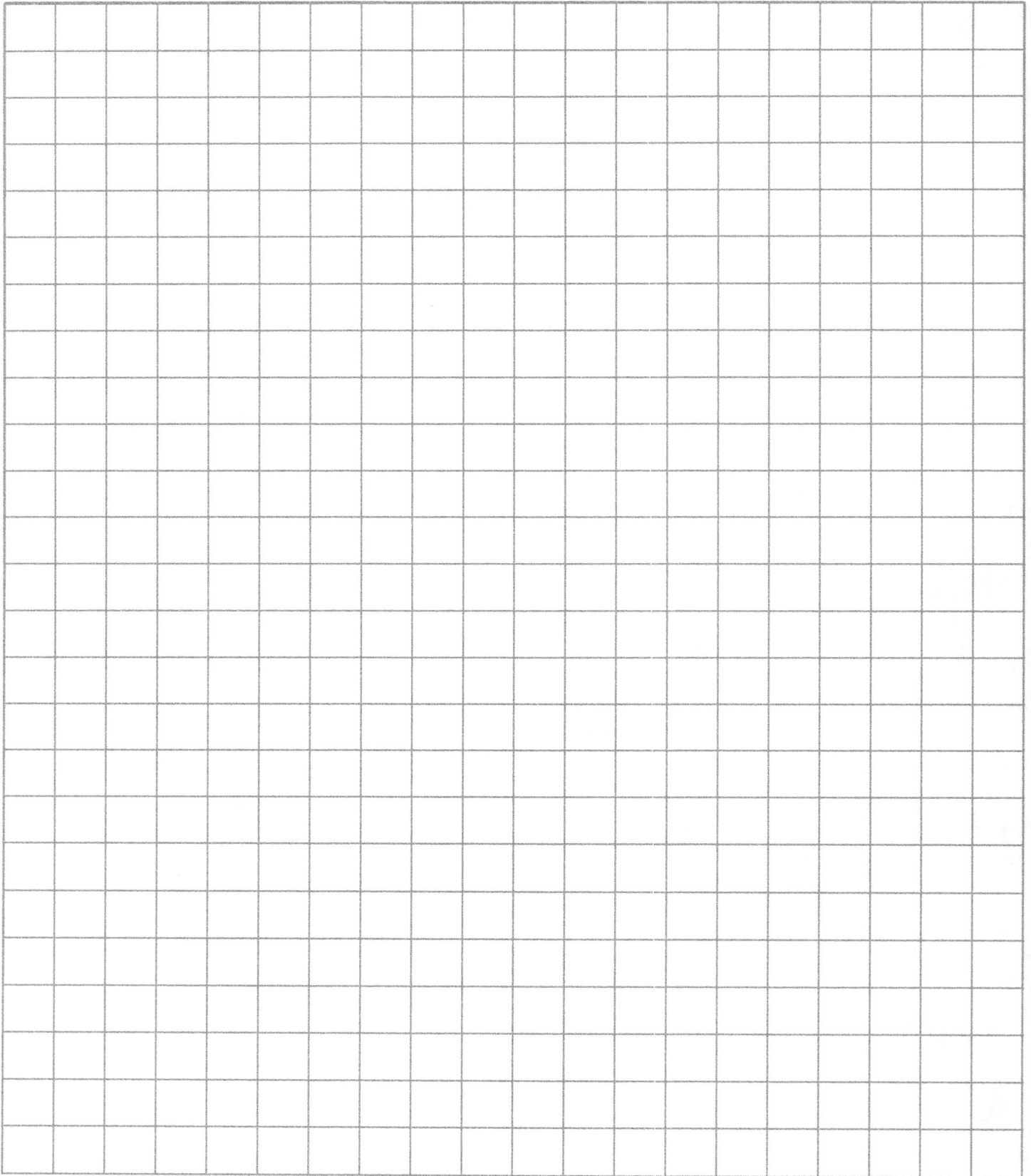

Blessings and Eternal Love

Graph Paper Deluxe

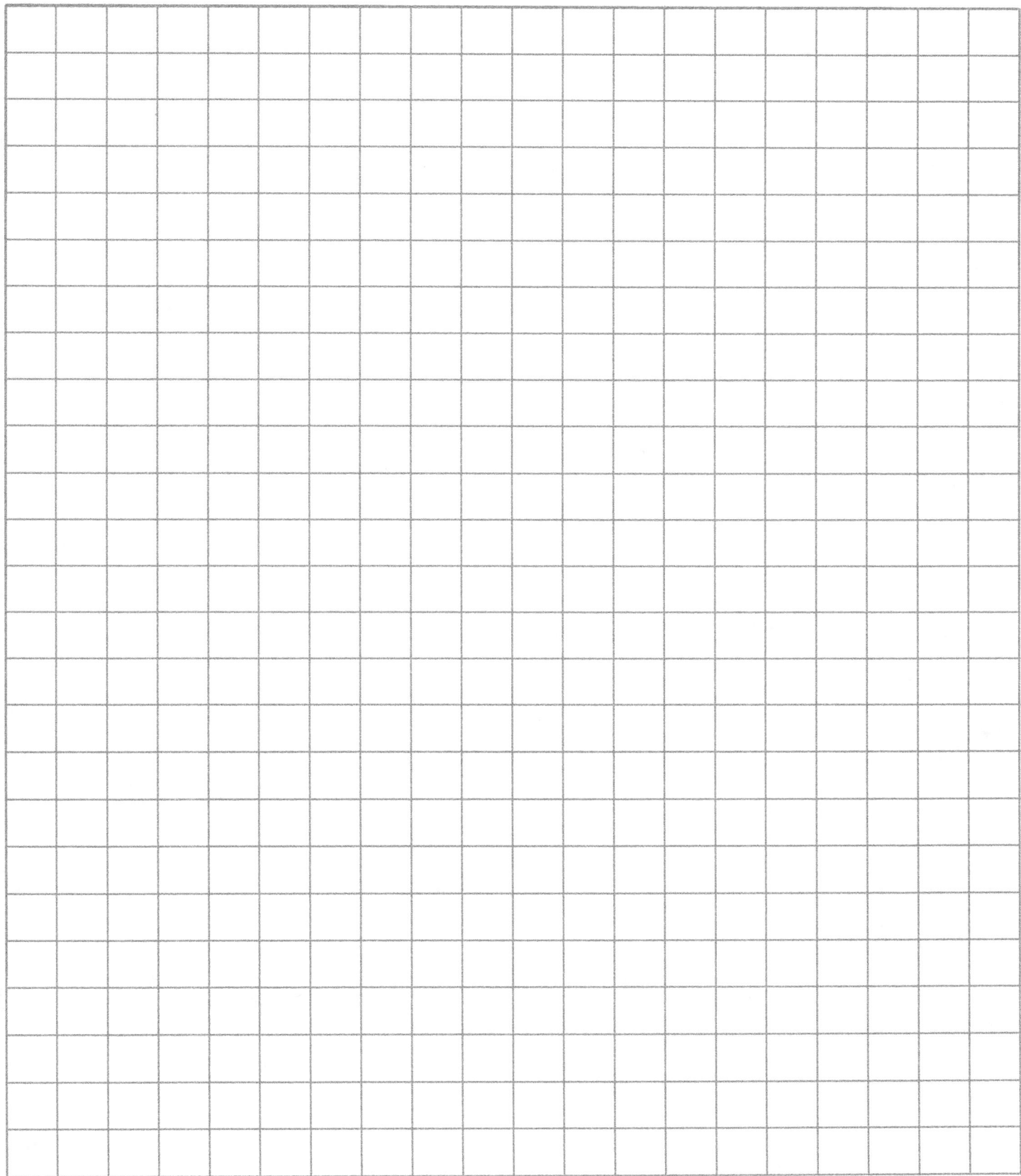

Kimberlite Kreations

Gray Line Edition

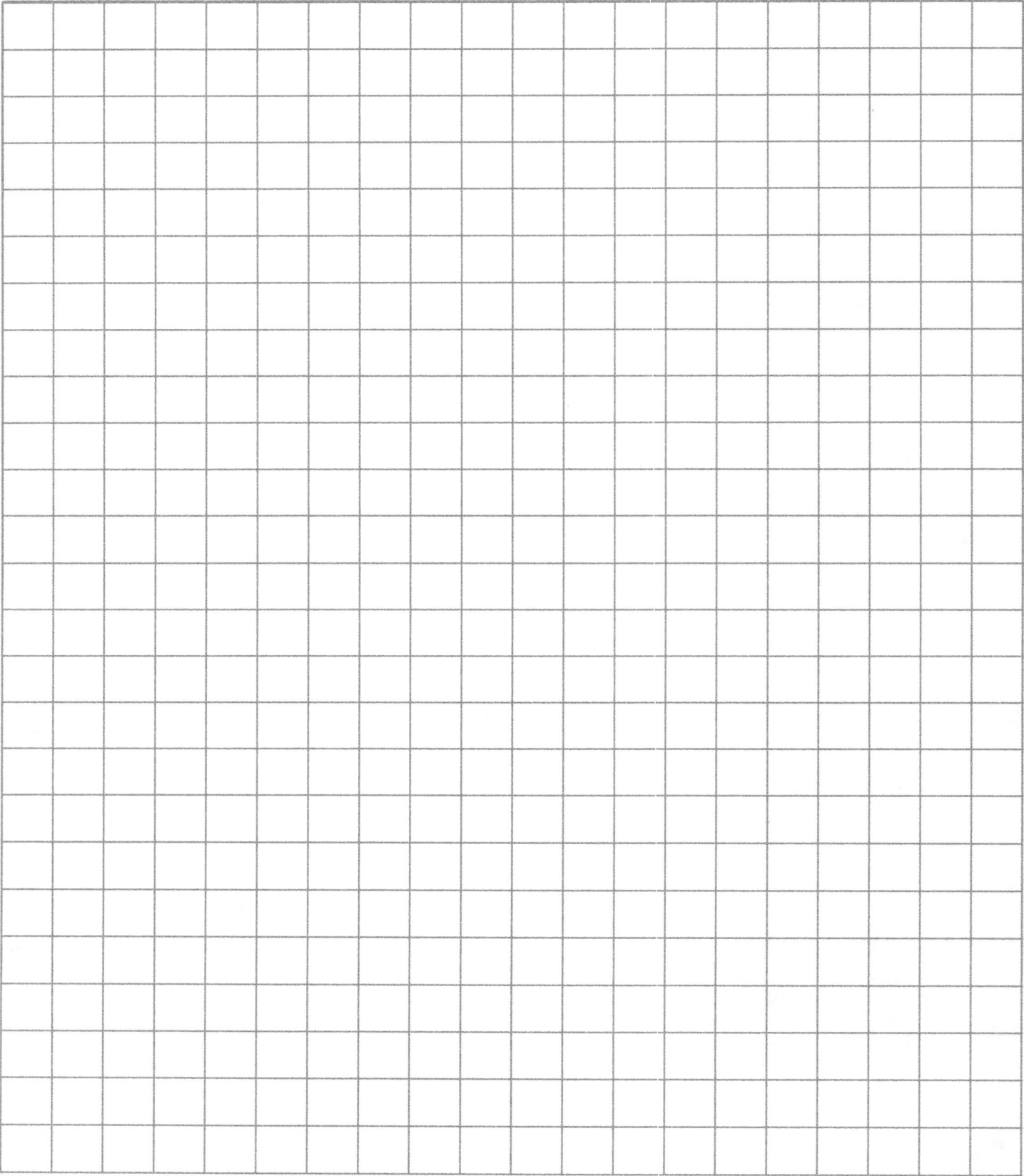

Blessings and Eternal Love

Graph Paper Deluxe

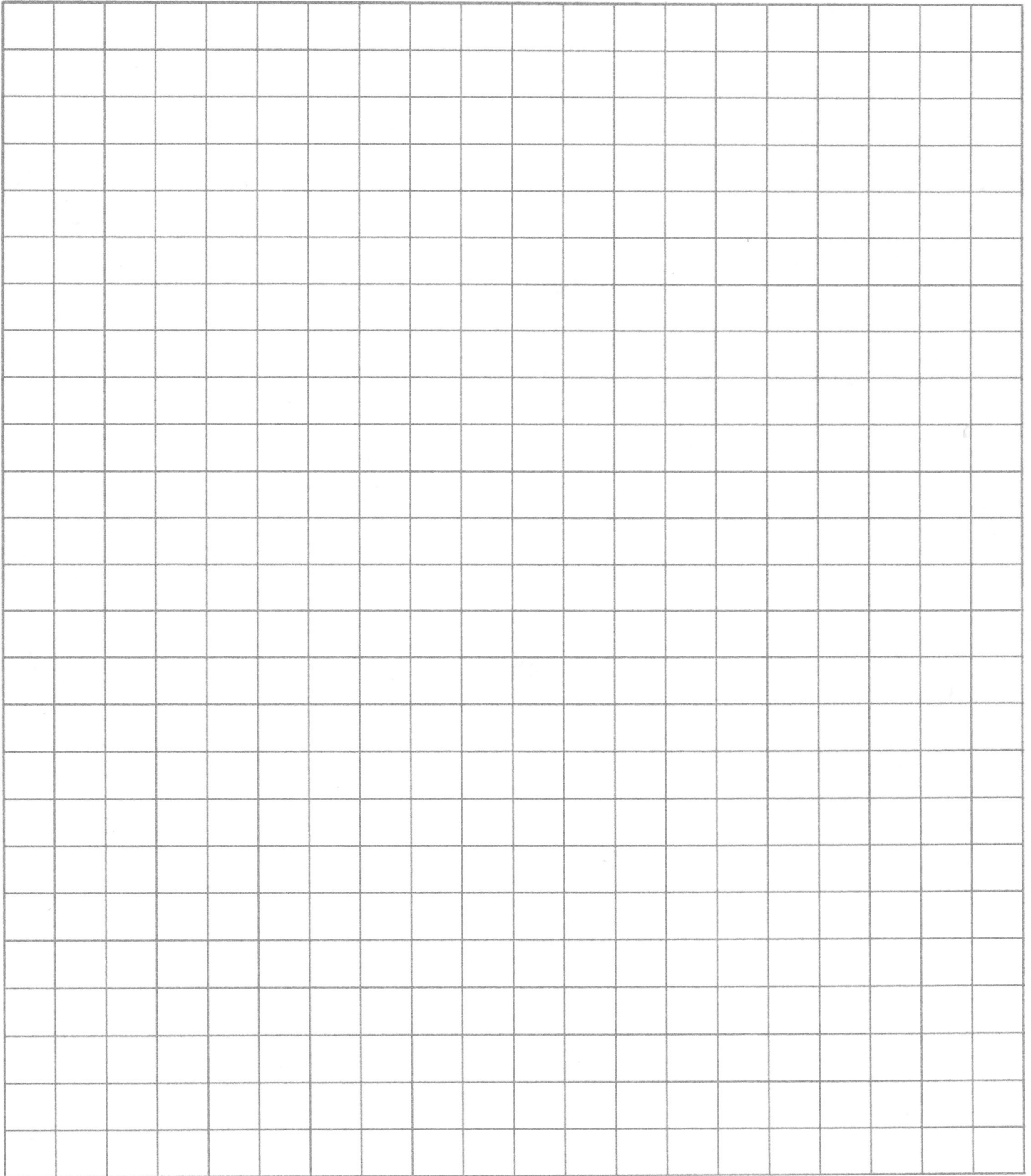

Gray Line Edition

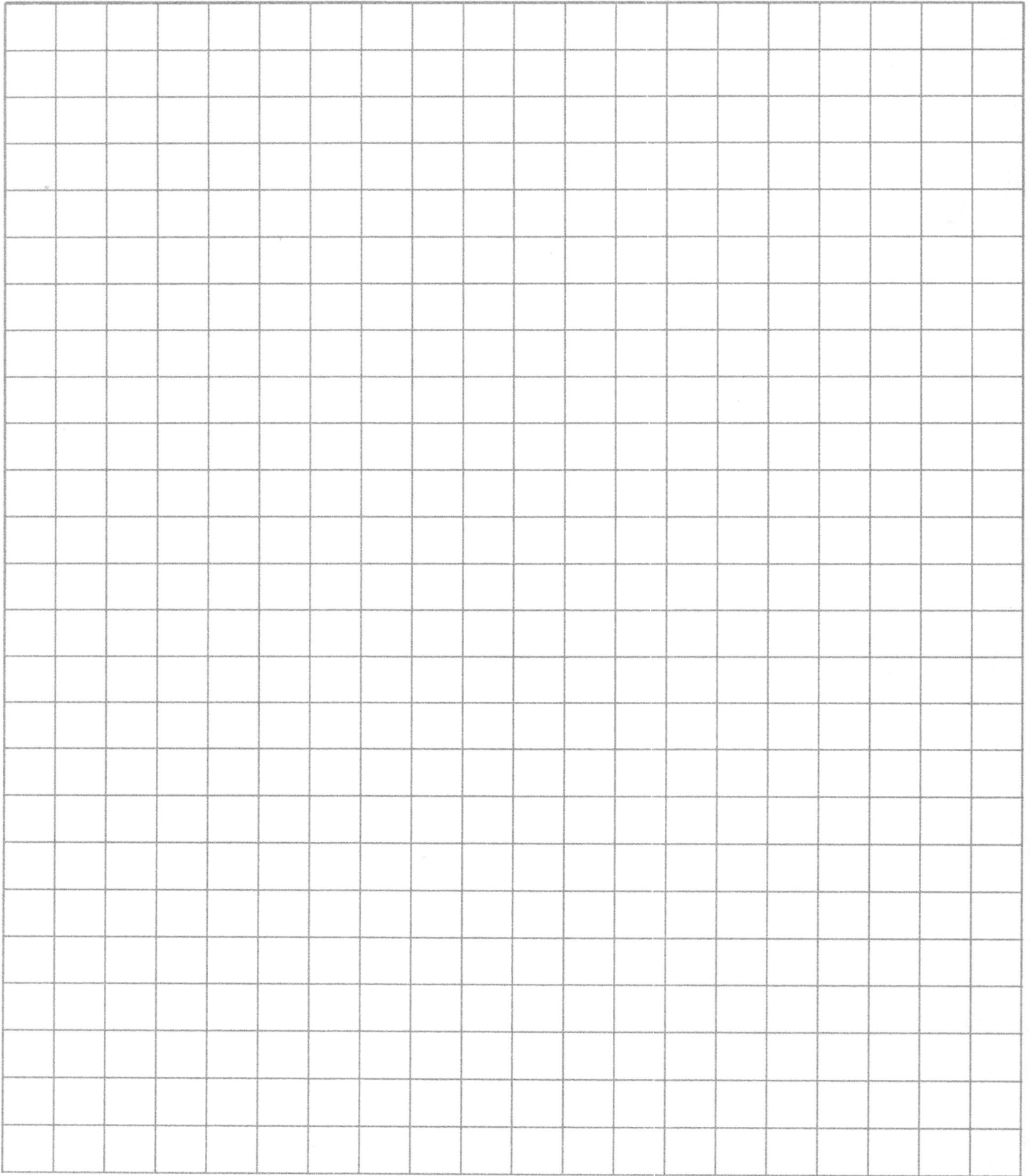

Blessings and Eternal Love

Graph Paper Deluxe

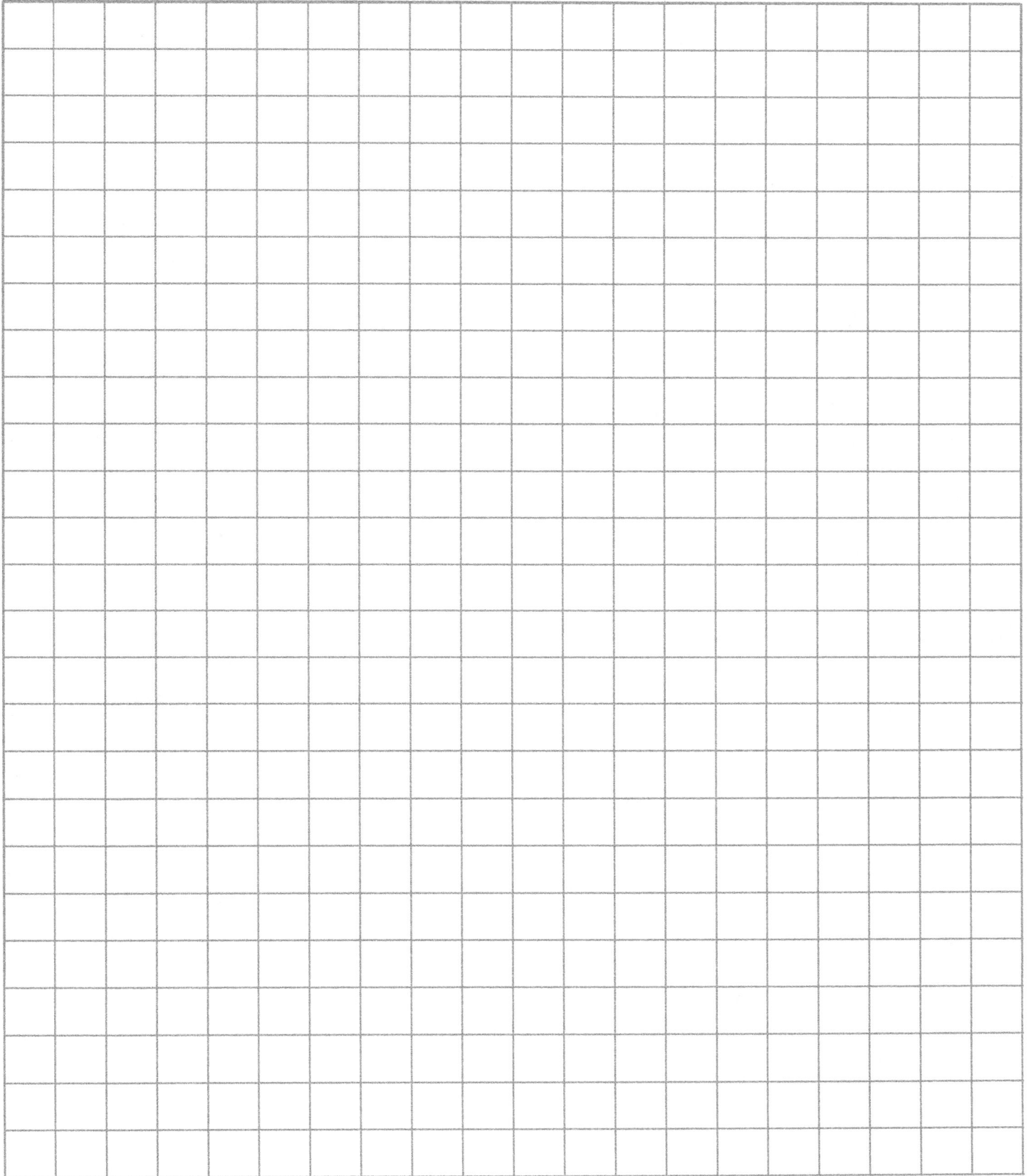

Gray Line Edition

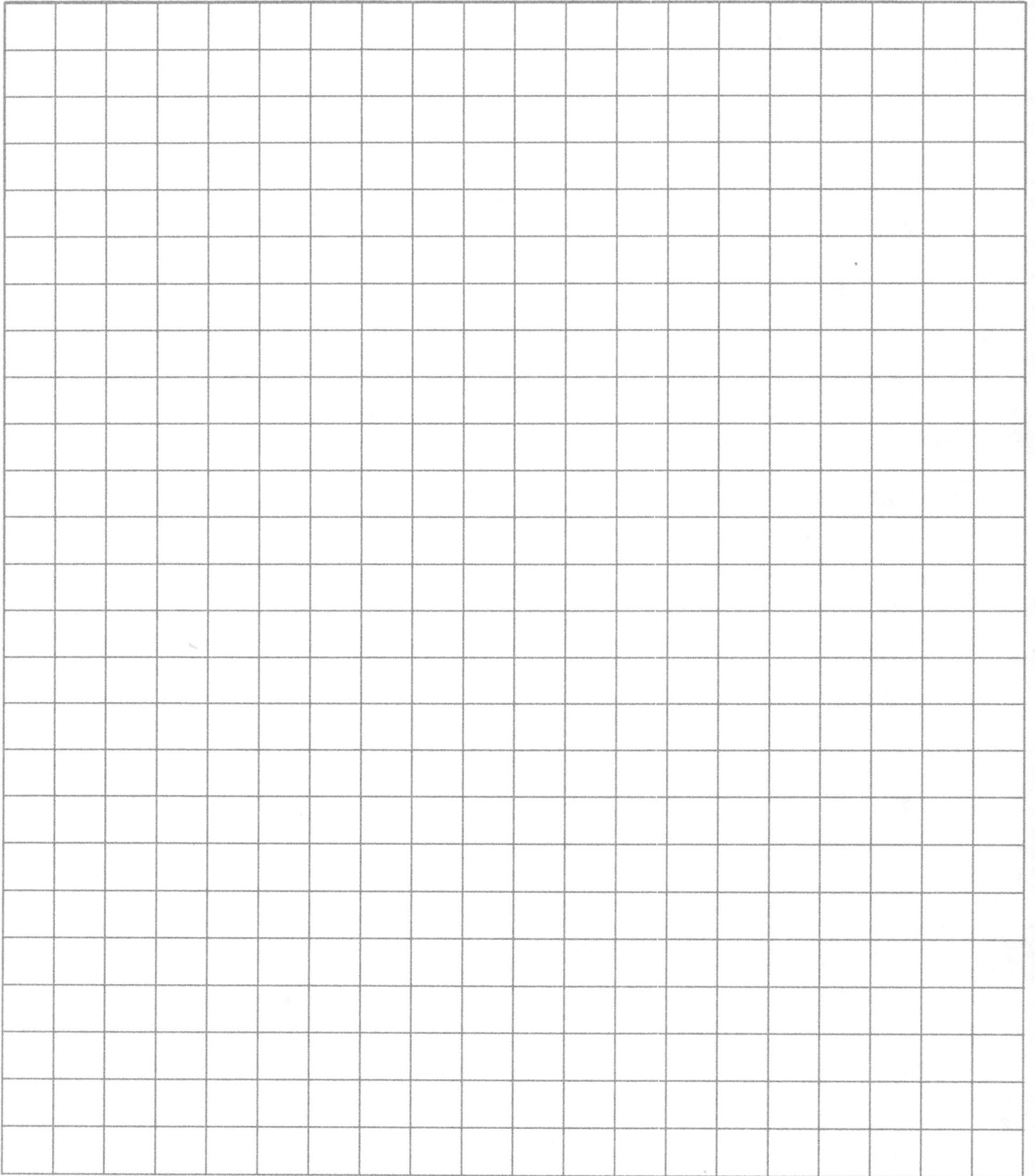

Blessings and Eternal Love

Graph Paper Deluxe

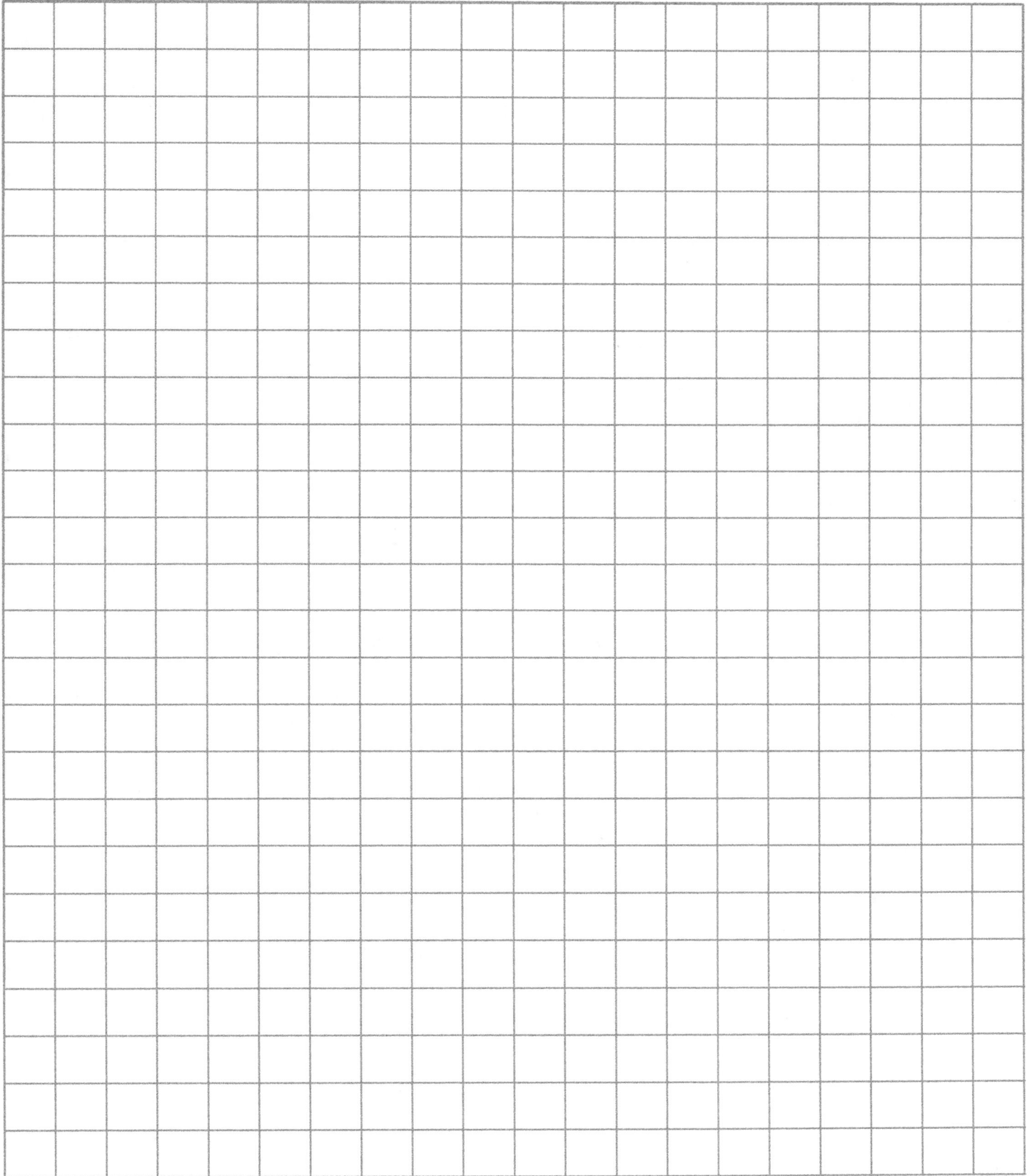

Gray Line Edition

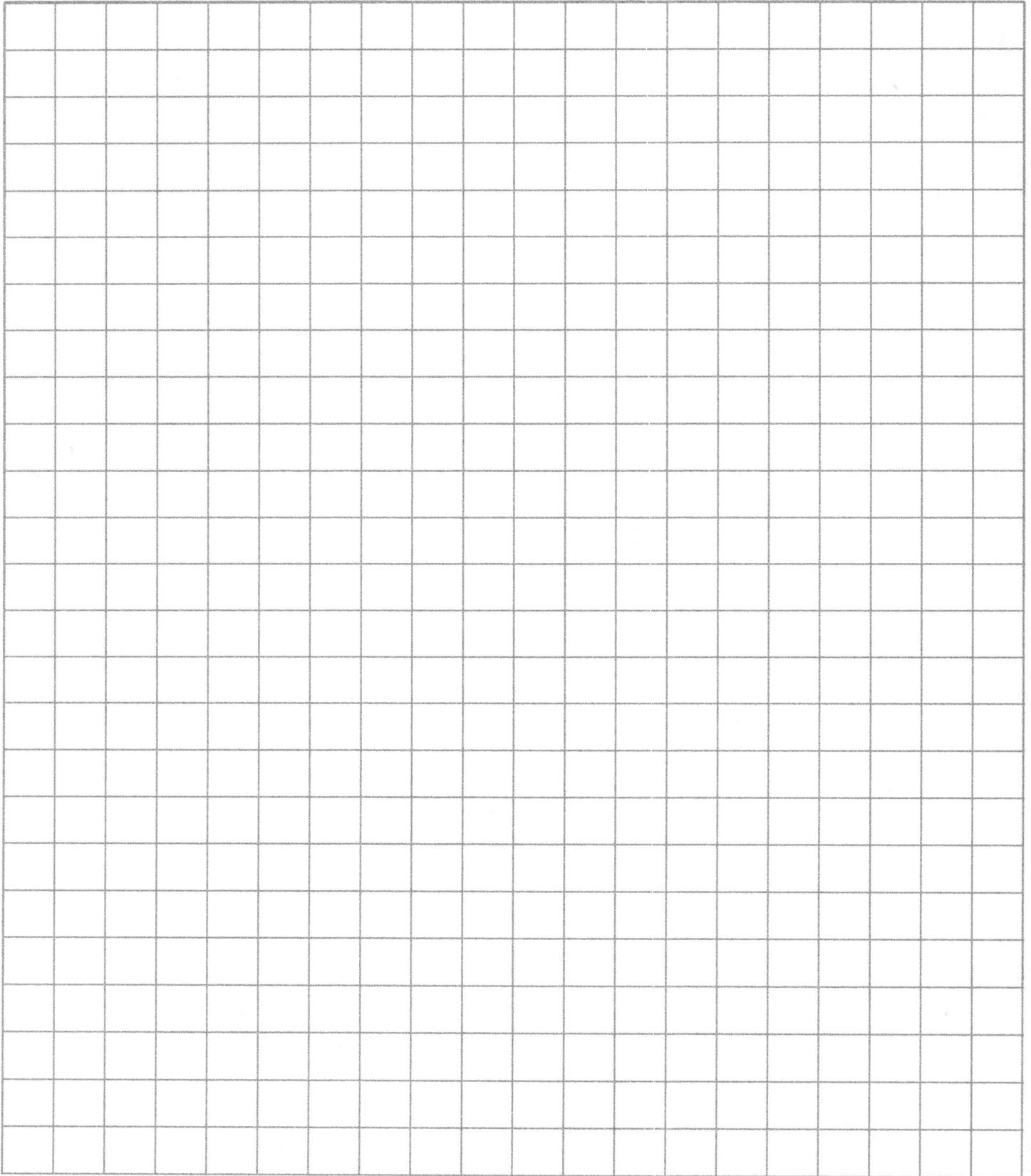

Blessings and Eternal Love

Graph Paper Deluxe

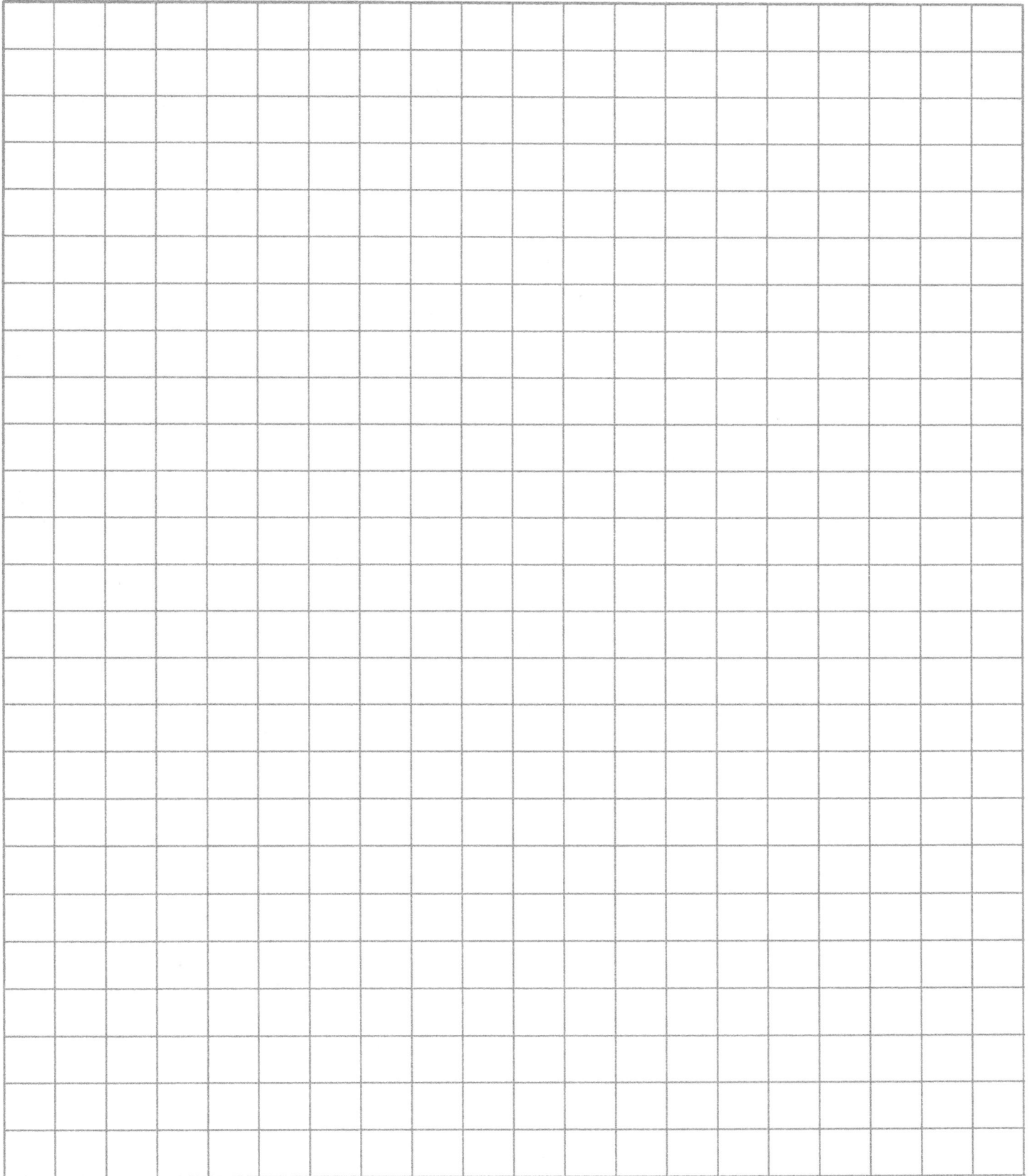

Gray Line Edition

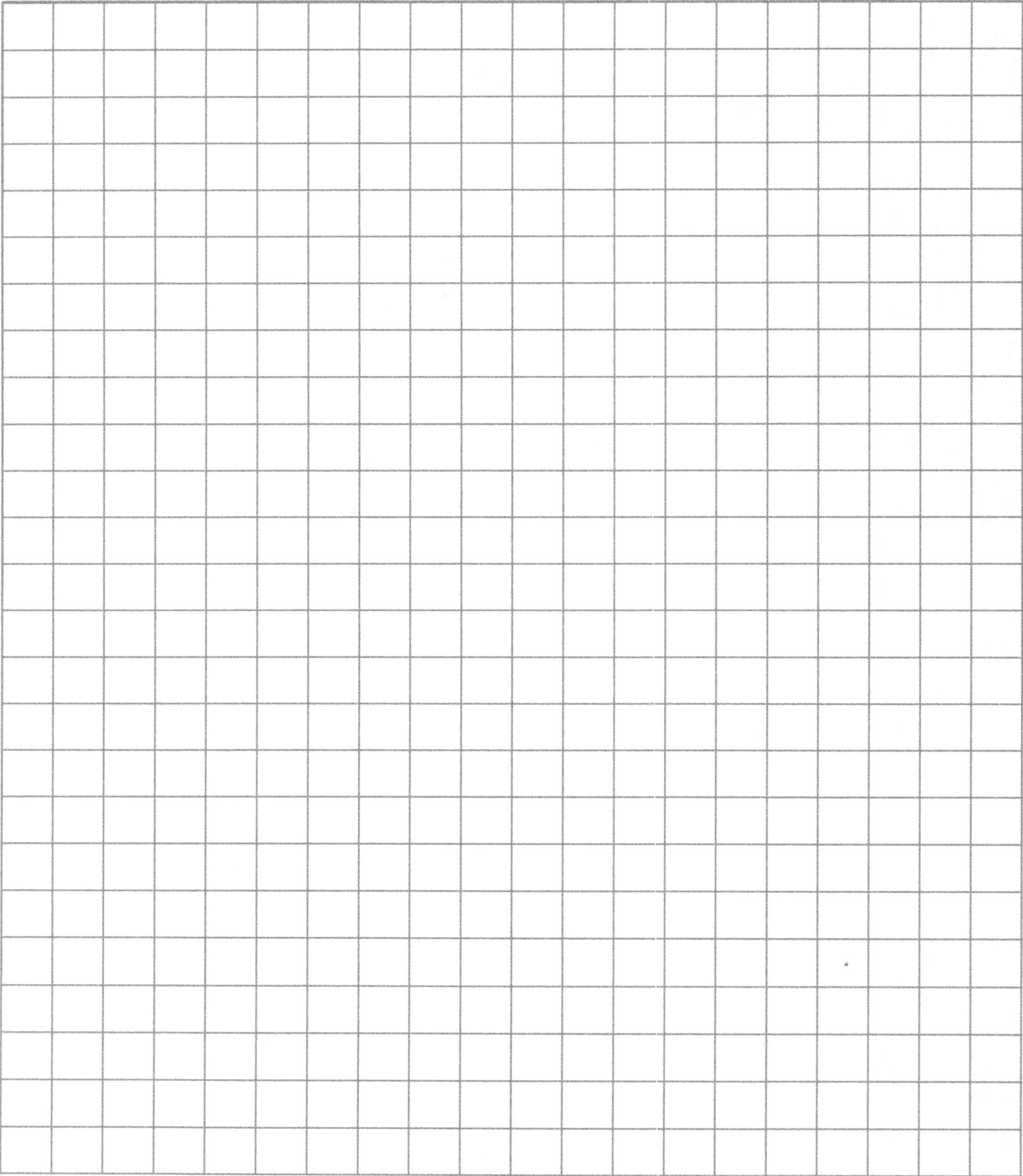

Blessings and Eternal Love

Graph Paper Deluxe

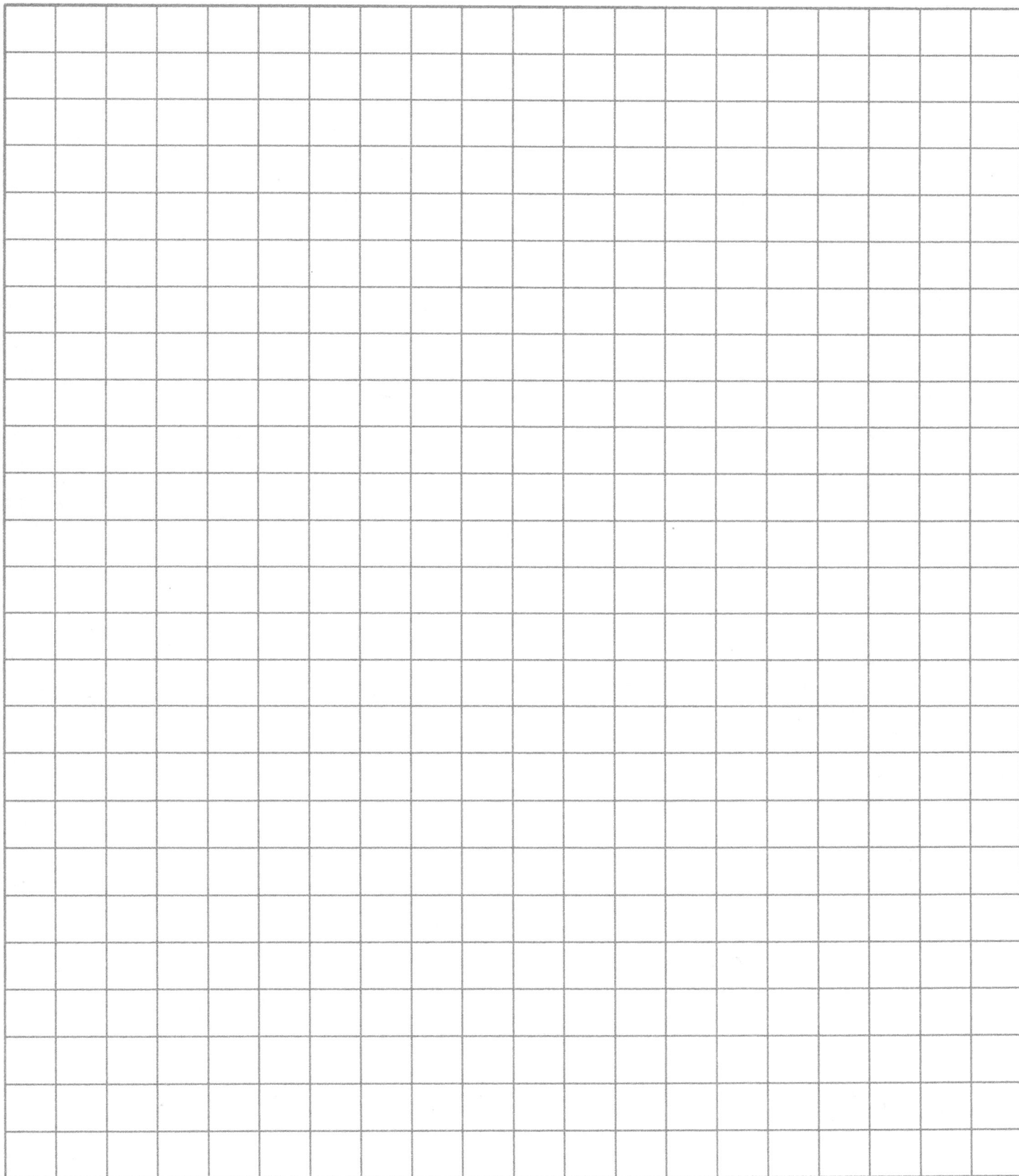

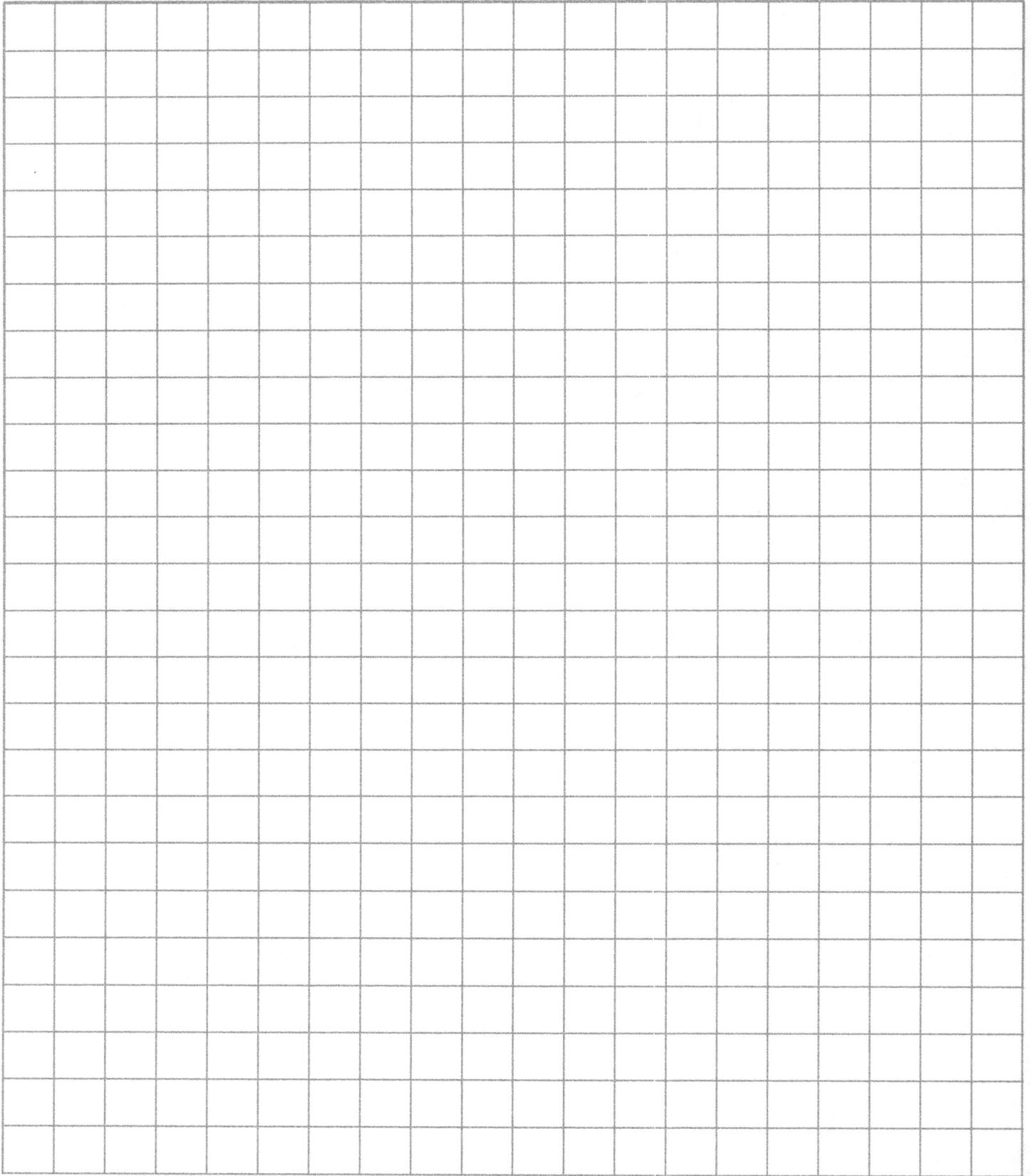

Graph Paper Deluxe

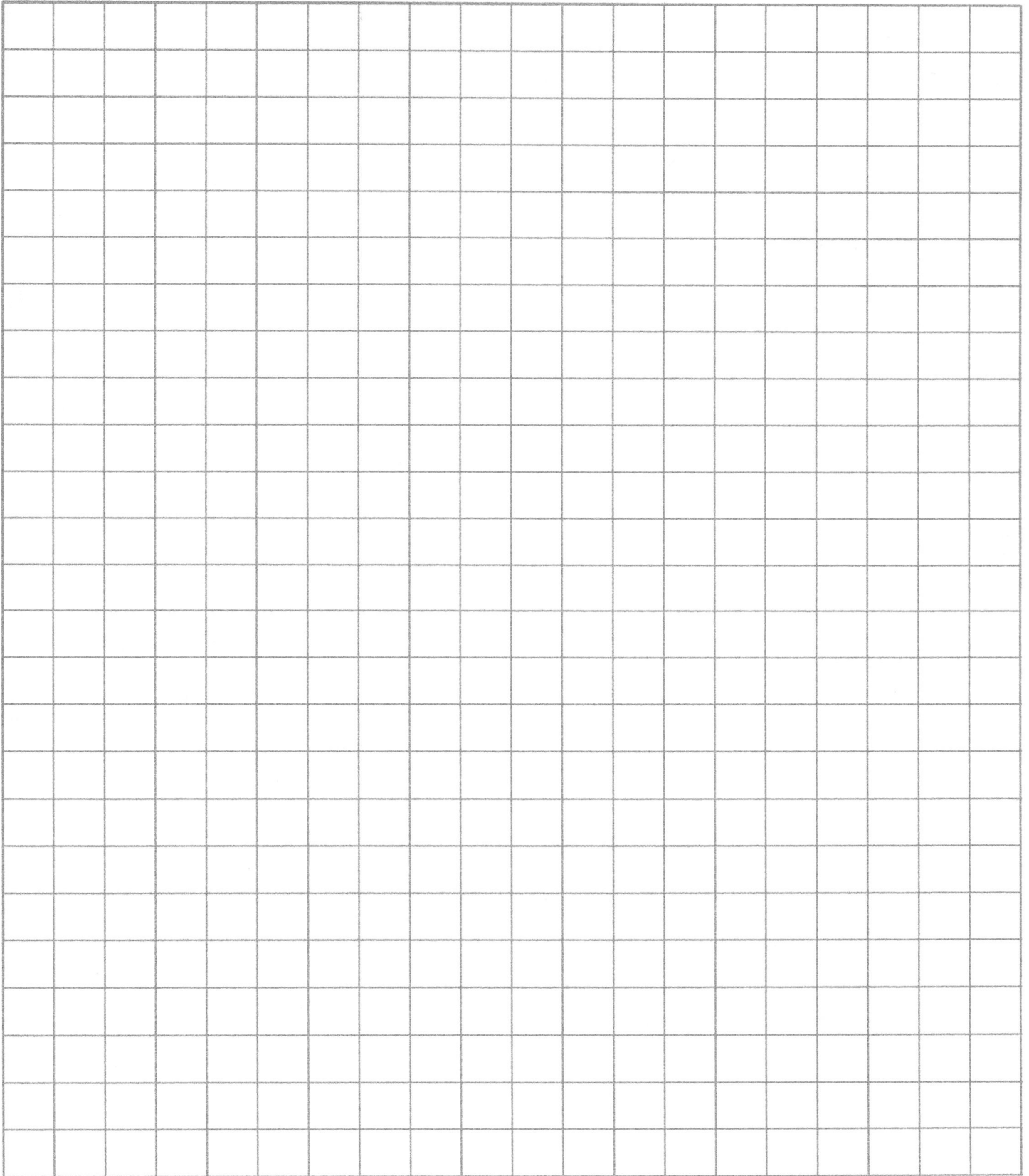

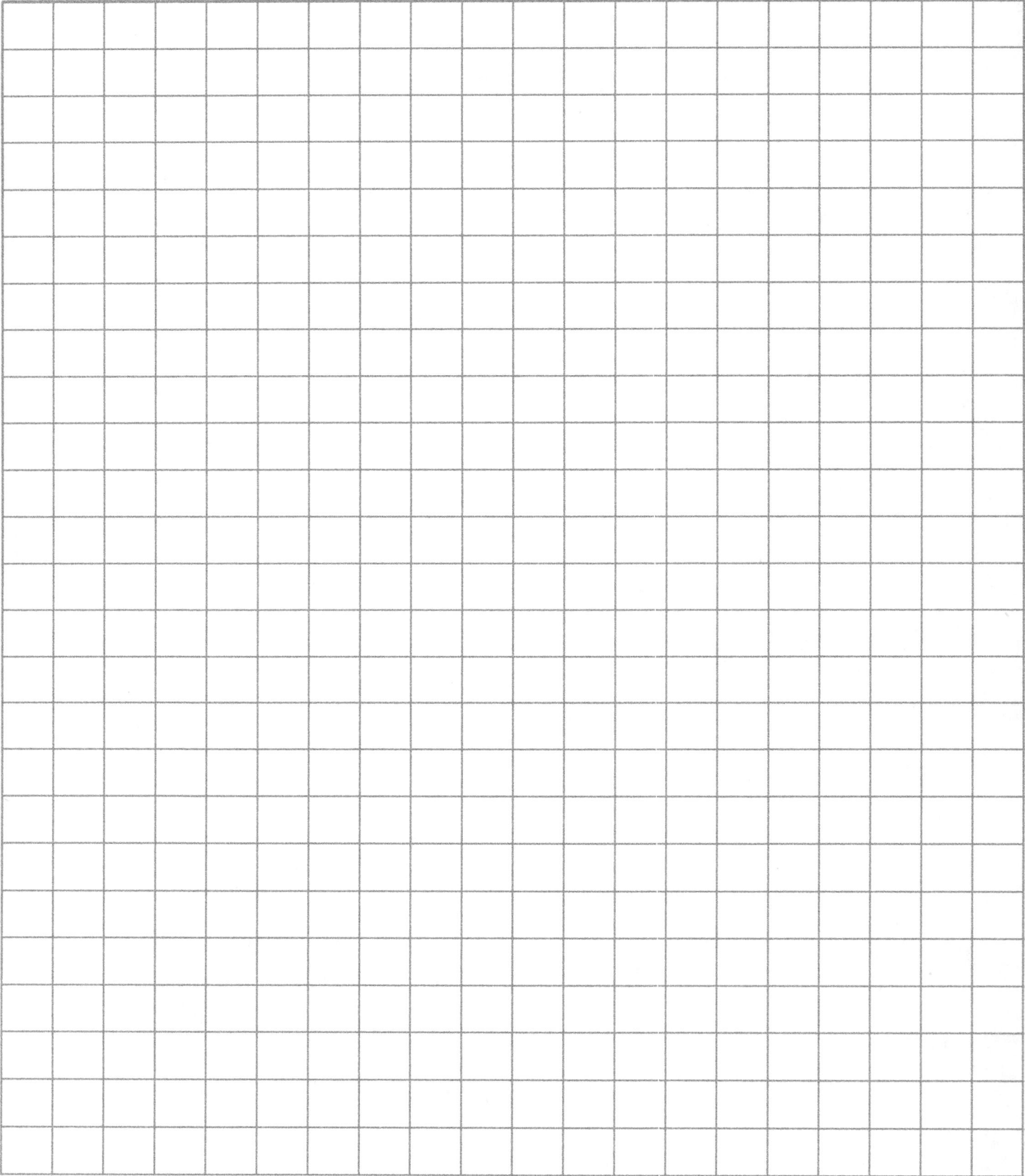

Graph Paper Deluxe

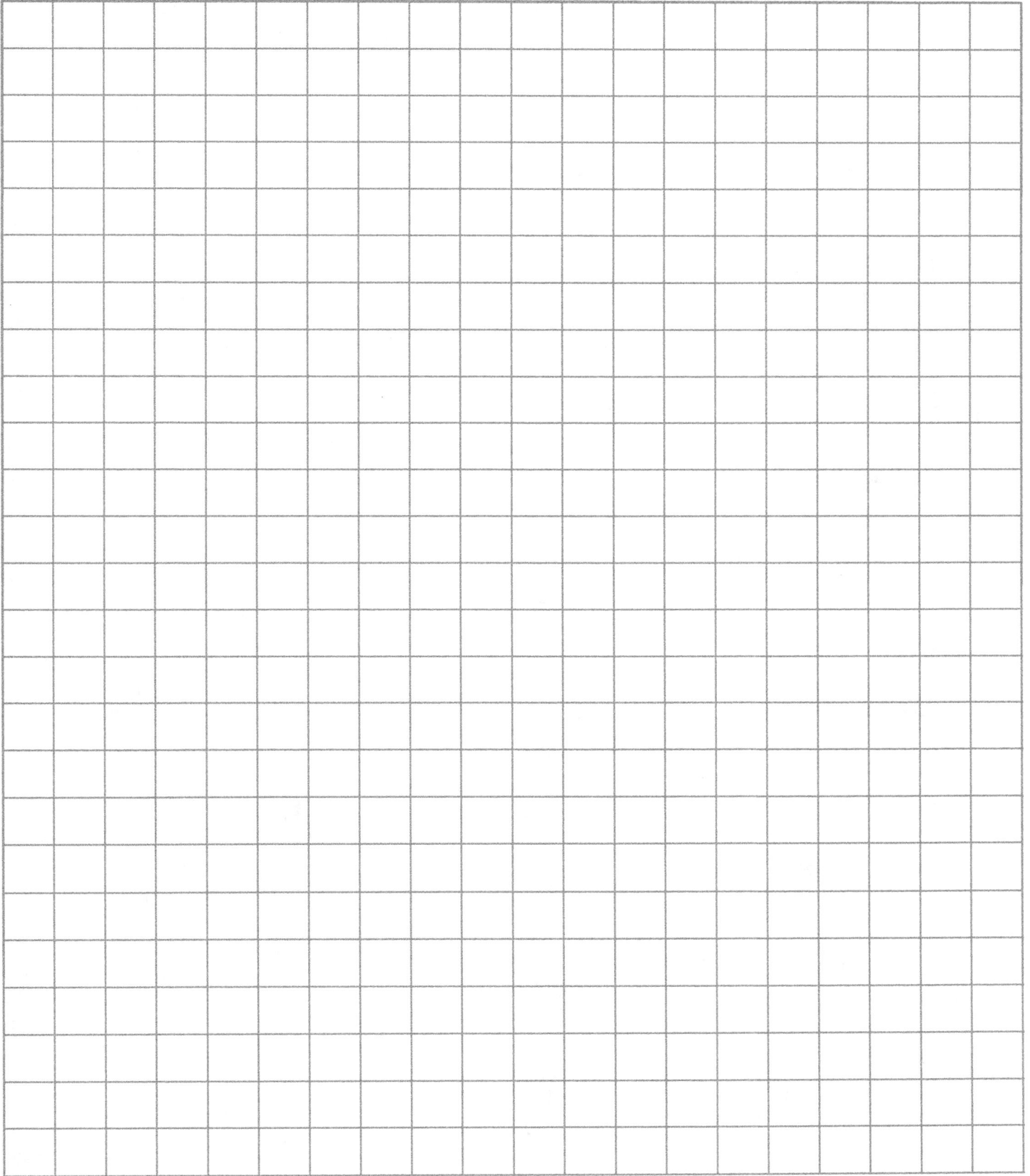

Graph Paper Deluxe

Kimberlite Kreations

Graph Paper Deluxe

Kimberlite Kreations

Graph Paper Deluxe

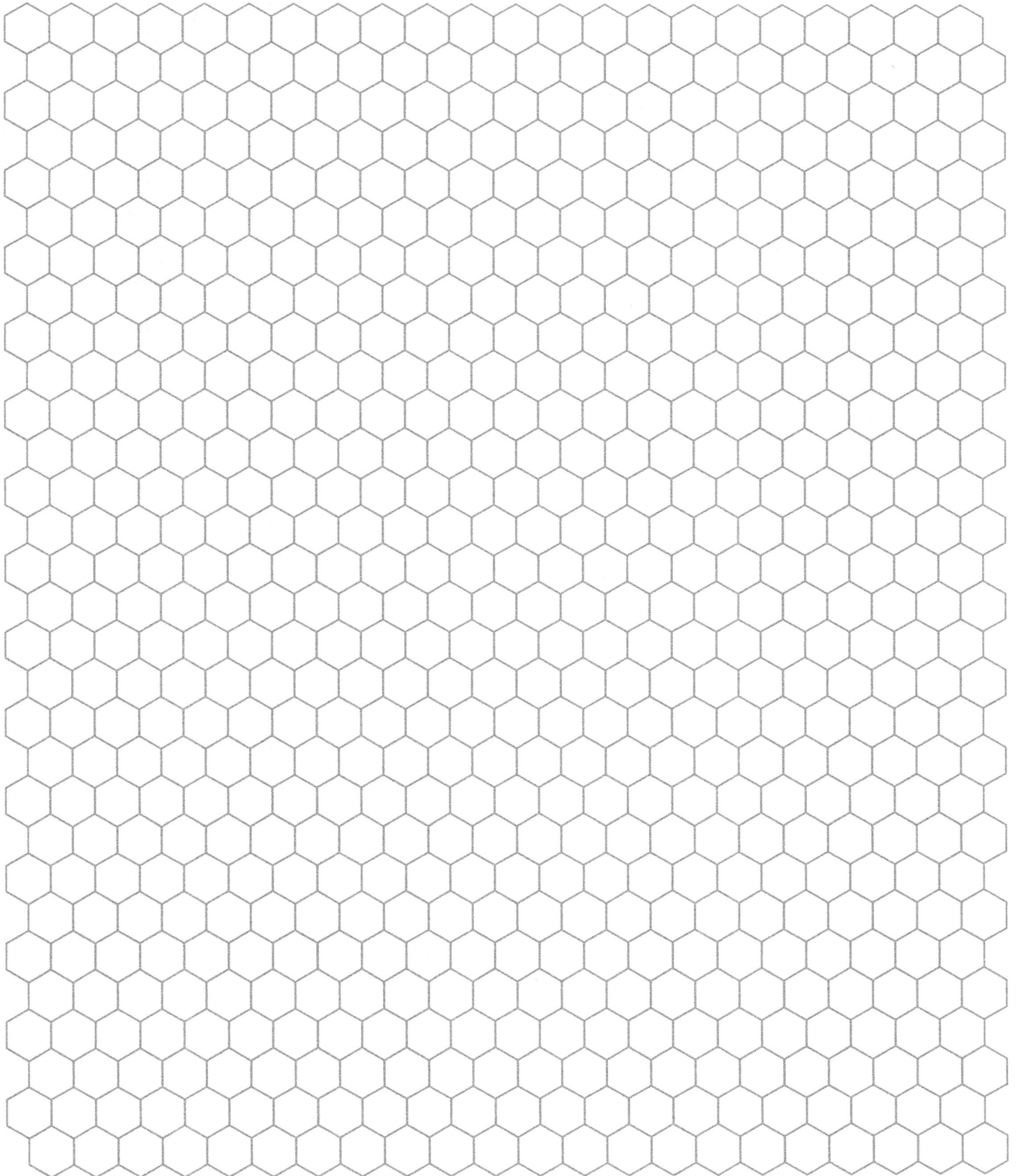

Kimberlite Kreations

Gray Line Edition

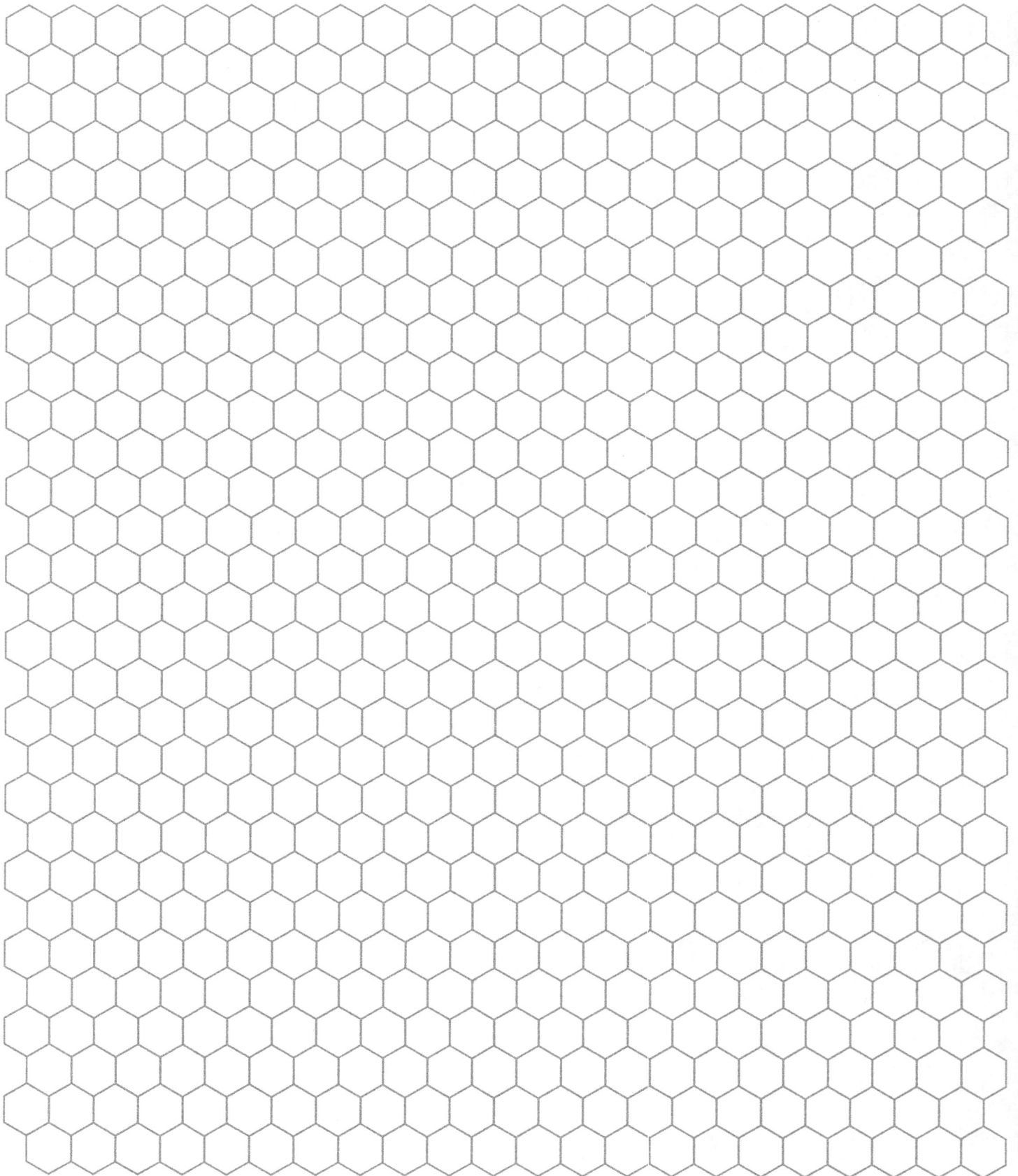

Blessings and Eternal Love

Graph Paper Deluxe

Graph Paper Deluxe

Gray Line Edition

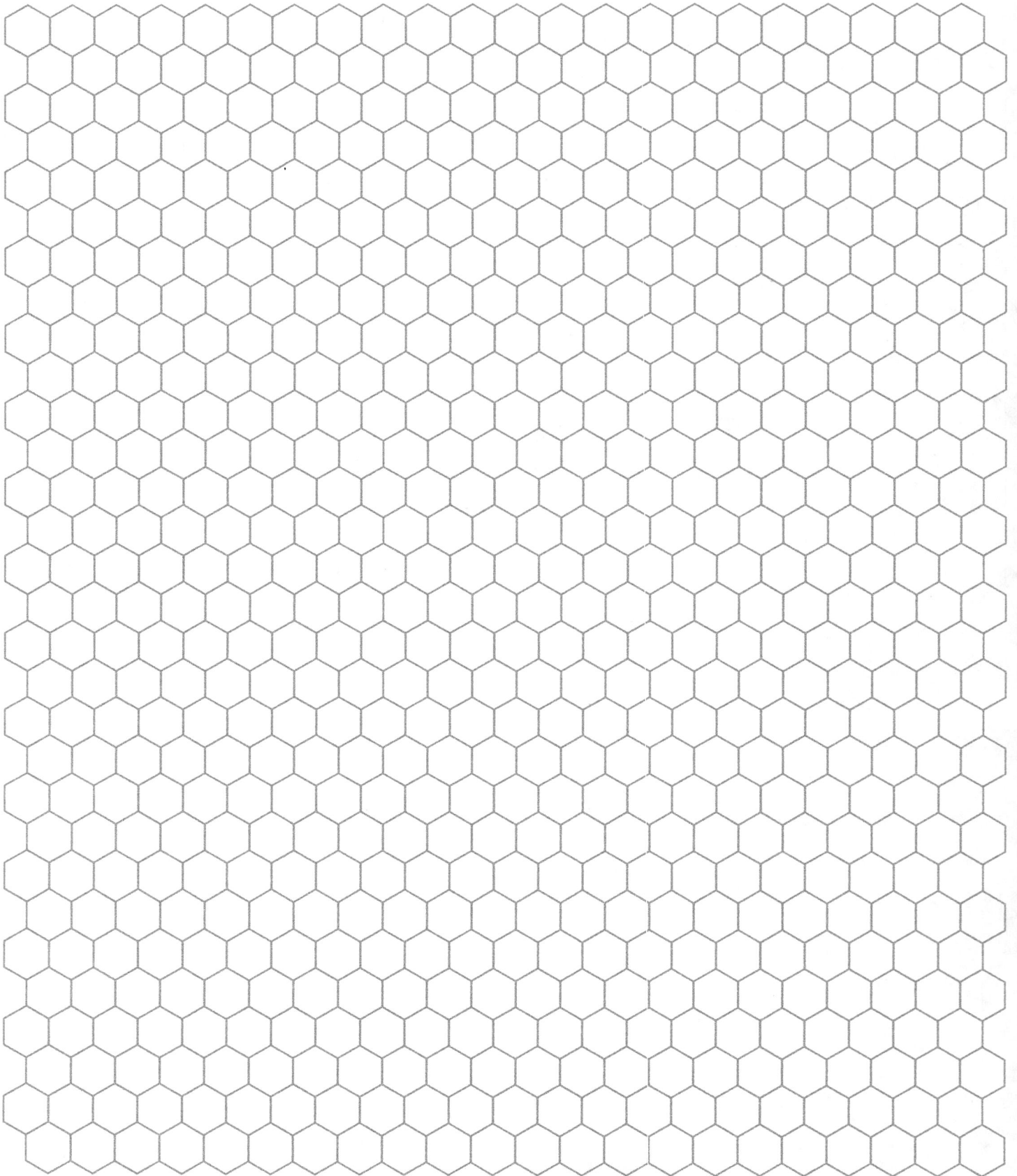

Blessings and Eternal Love

Graph Paper Deluxe

Graph Paper Deluxe

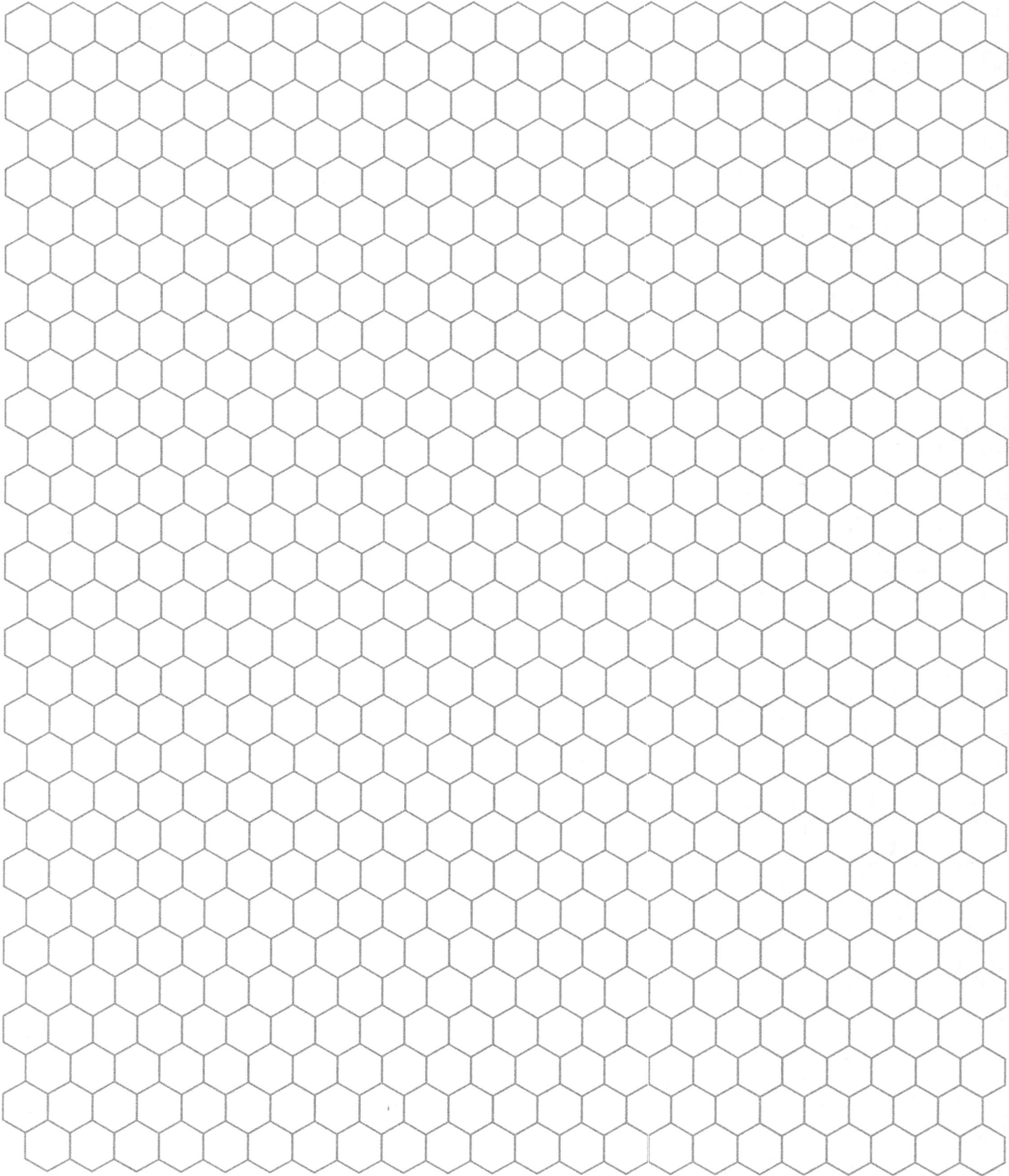

Blessings and Eternal Love

Graph Paper Deluxe

Kimberlite Kreations

Graph Paper Deluxe

Graph Paper Deluxe

Graph Paper Deluxe

Gray Line Edition

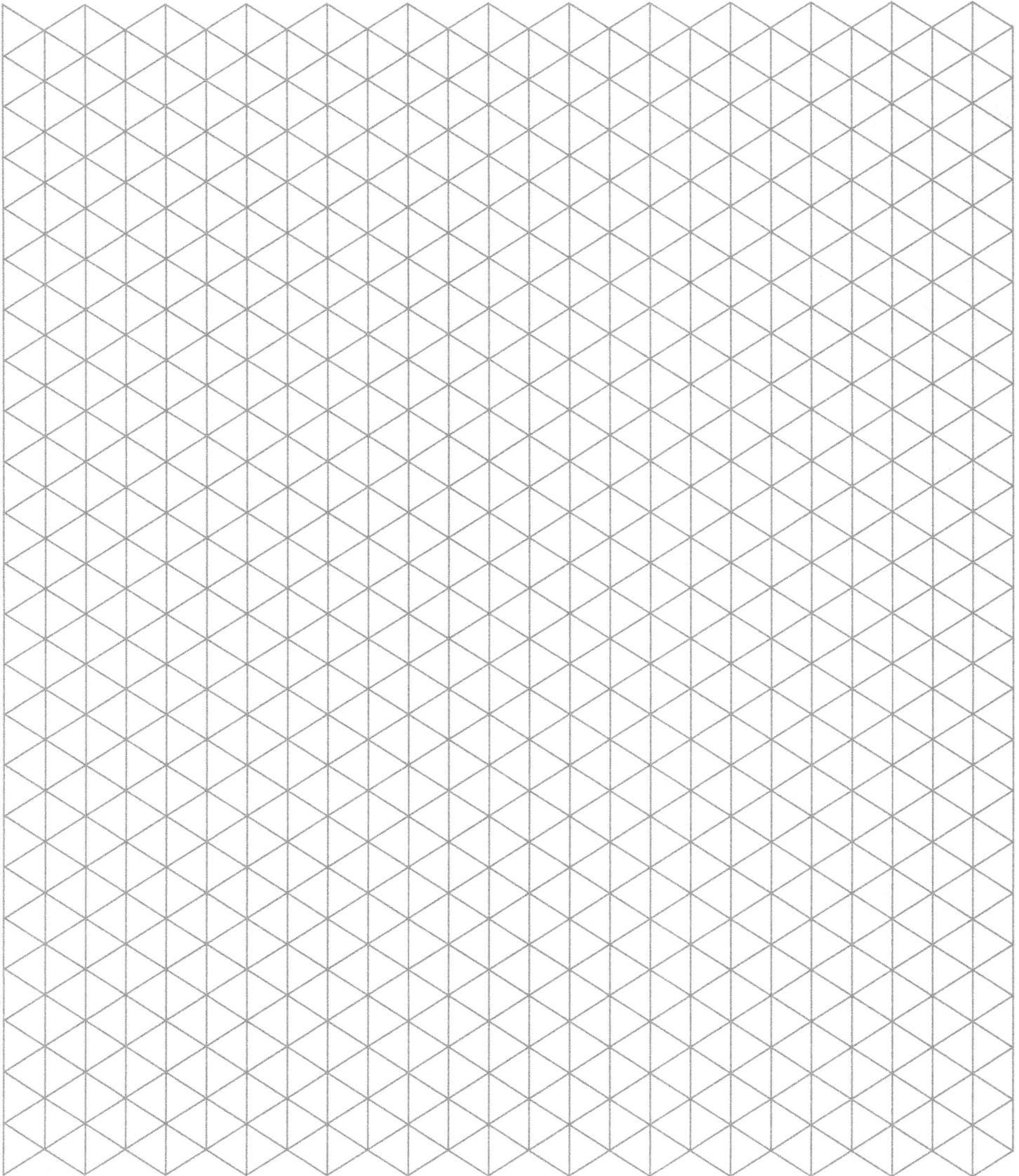

Blessings and Eternal Love

Graph Paper Deluxe

Gray Line Edition

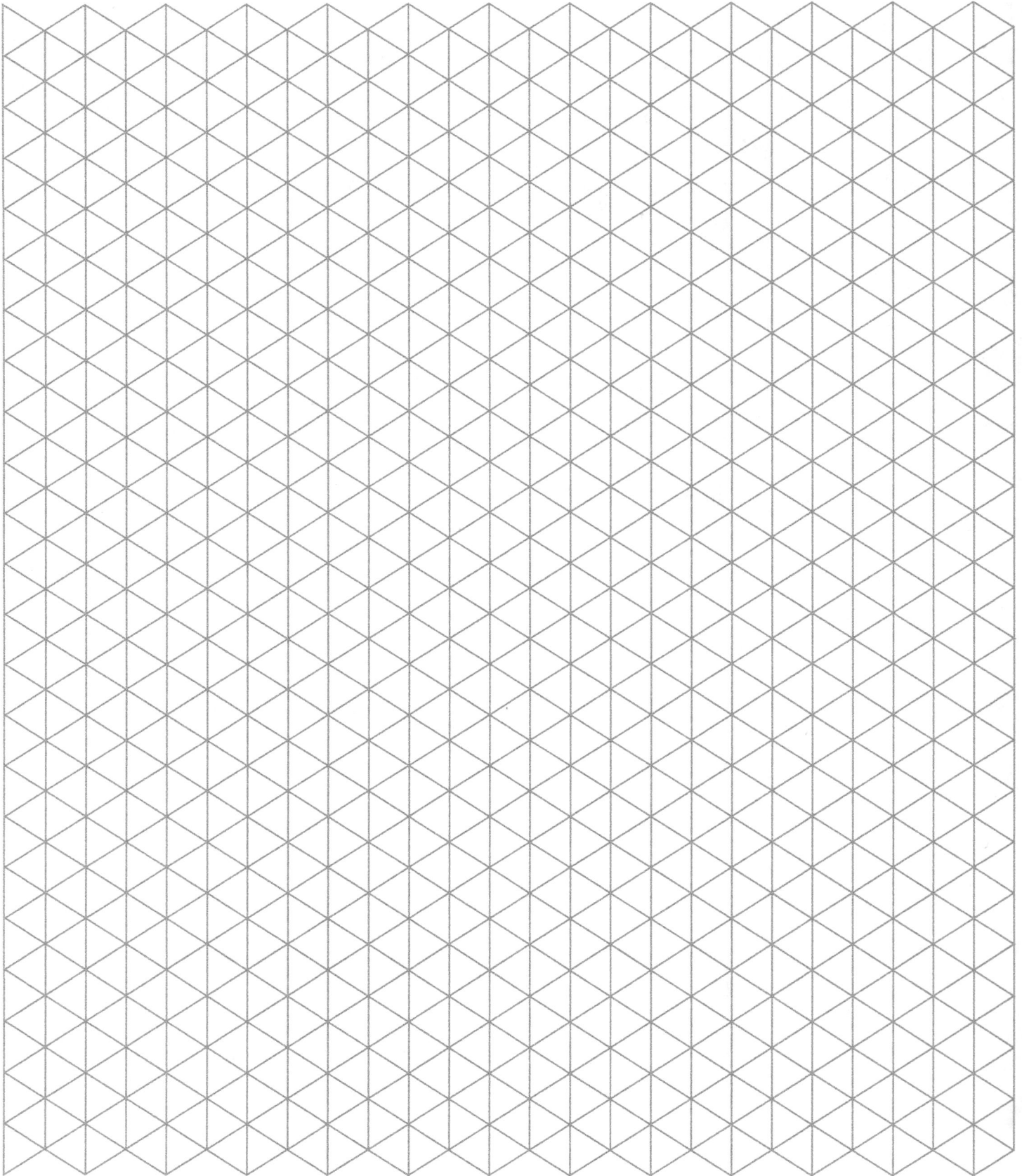

Blessings and Eternal Love

Graph Paper Deluxe

Graph Paper Deluxe

Blessings and Eternal Love

Graph Paper Deluxe

Gray Line Edition

Blessings and Eternal Love

Graph Paper Deluxe

Kimberlite Kreations

Gray Line Edition

Blessings and Eternal Love

Graph Paper Deluxe

Graph Paper Deluxe

Gray Line Edition

Blessings and Eternal Love

Graph Paper Deluxe

Gray Line Edition

Blessings and Eternal Love

Graph Paper Deluxe

Gray Line Edition

Blessings and Eternal Love

Graph Paper Deluxe

Graph Paper Deluxe

Graph Paper Deluxe

Graph Paper Deluxe

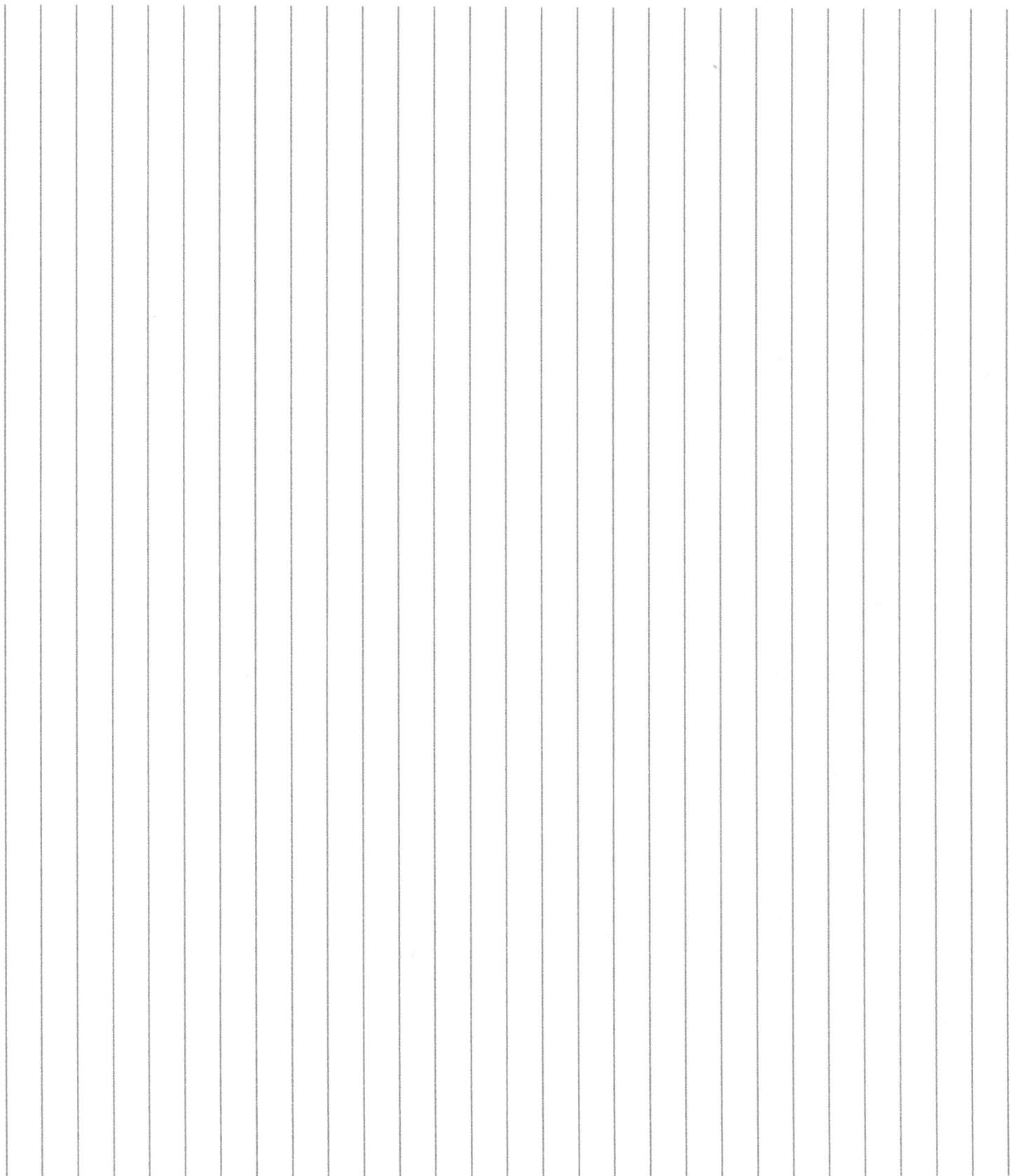

Kimberlite Kreations

Graph Paper Deluxe

Graph Paper Deluxe

Graph Paper Deluxe

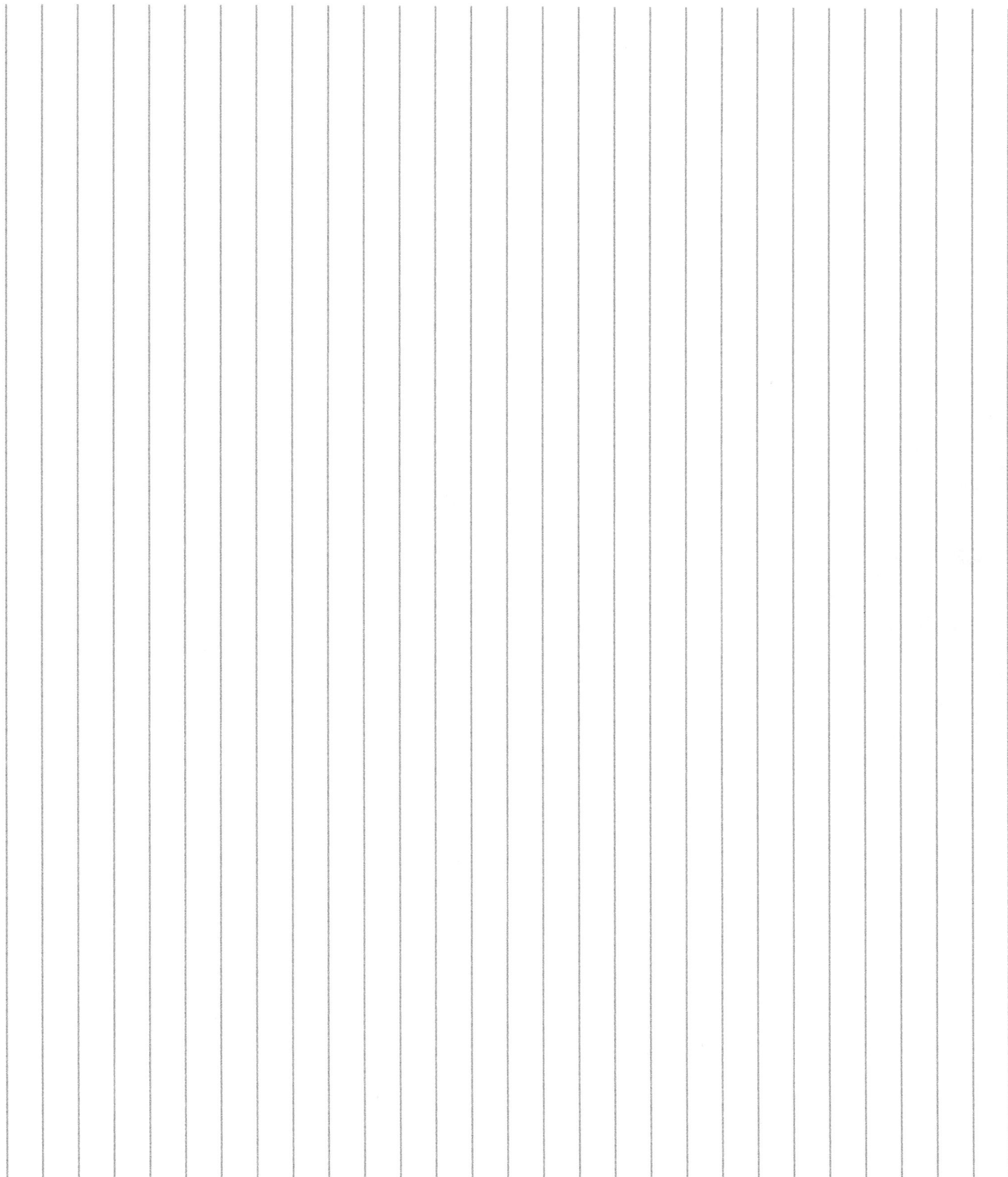

Gray Line Edition

Blessings and Eternal Love

Graph Paper Deluxe

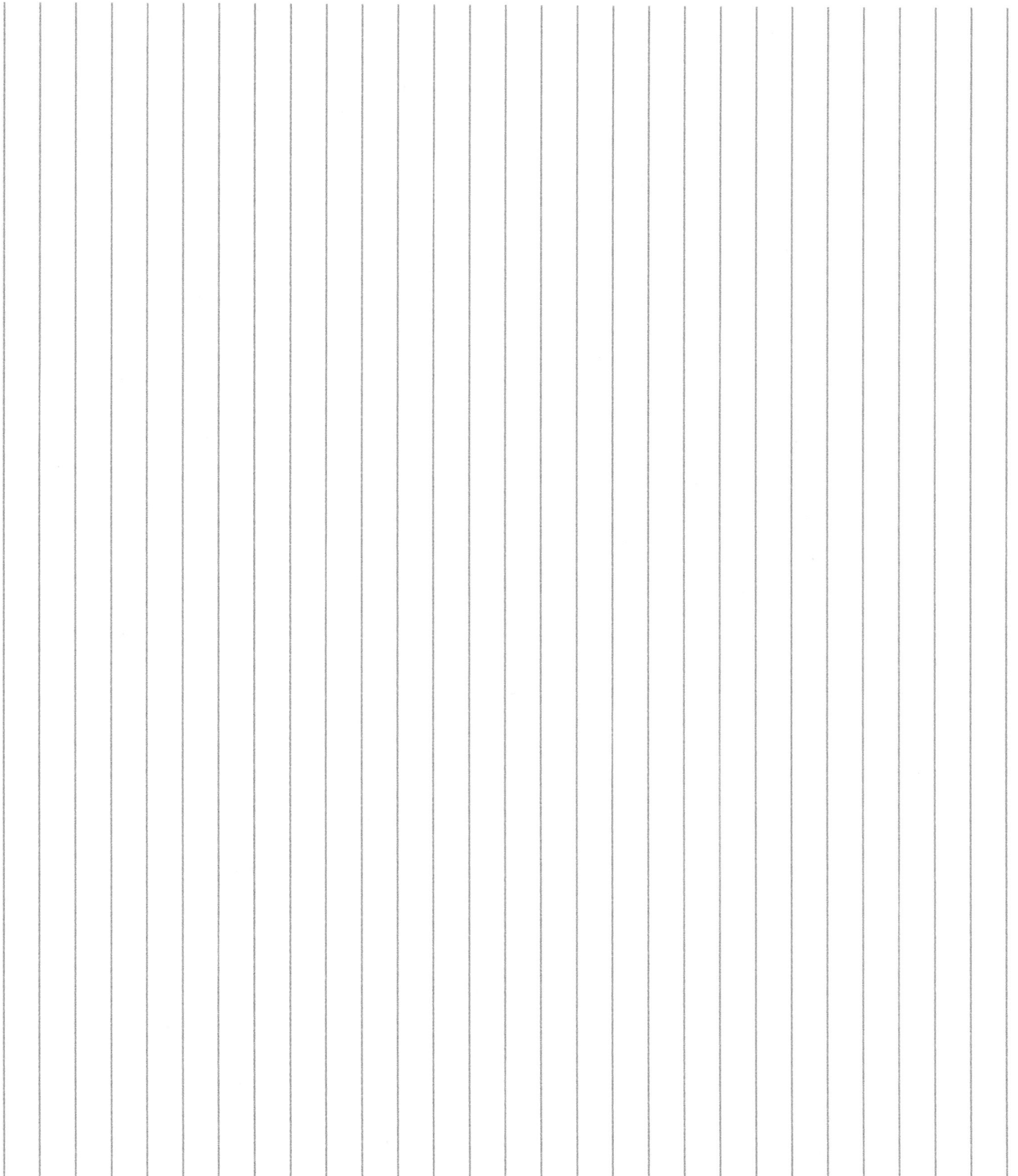

Gray Line Edition

Blessings and Eternal Love

Graph Paper Deluxe

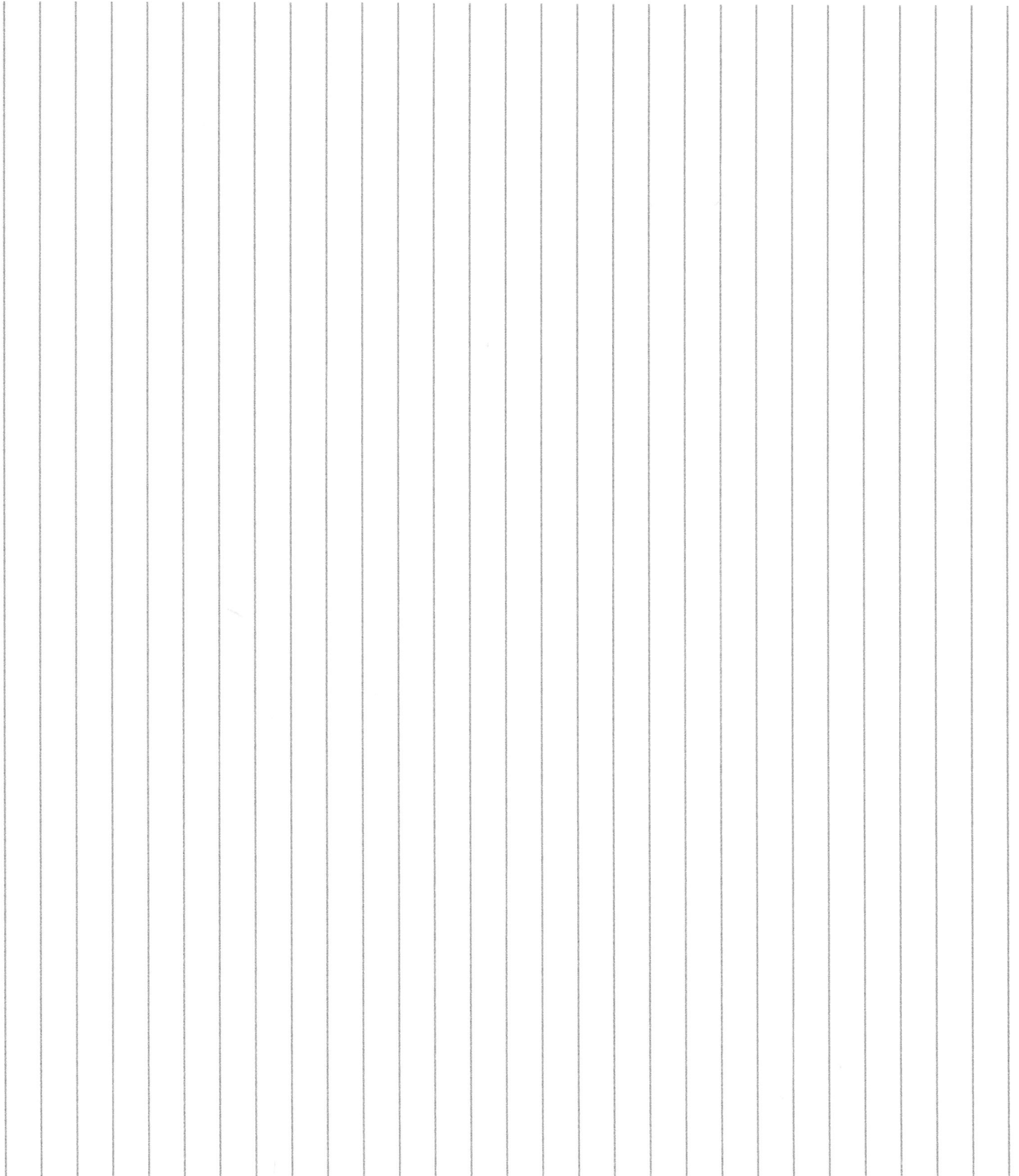

Gray Line Edition

Blessings and Eternal Love

Graph Paper Deluxe

Graph Paper Deluxe

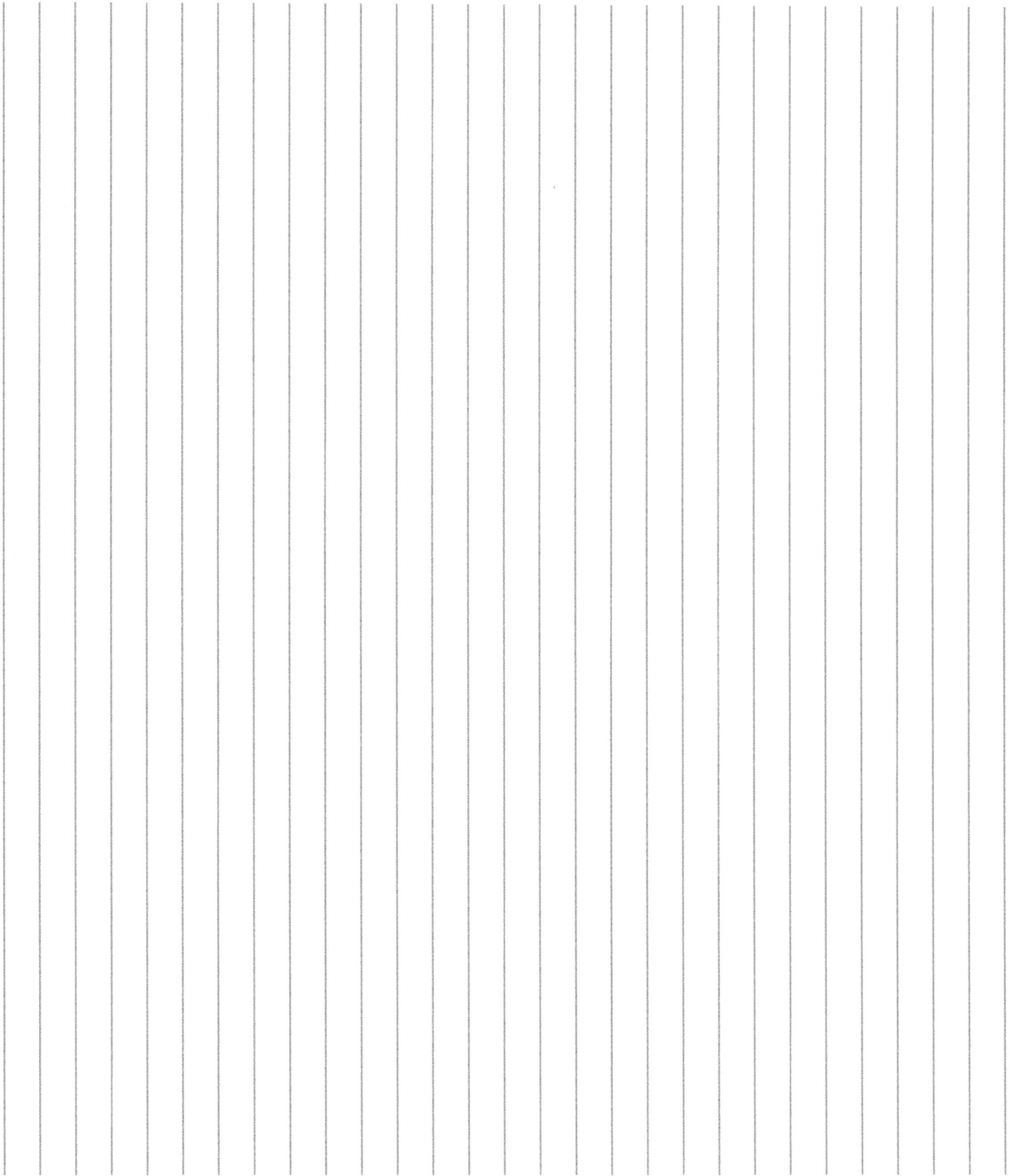

Gray Line Edition

Blessings and Eternal Love

Graph Paper Deluxe